Reviews of Environmental Contamination and Toxicology

VOLUME 163

Springer
New York
Berlin
Heidelberg
Barcelona
Hong Kong
London
Milan
Paris
Singapore
Tokyo

Reviews of Environmental Contamination and Toxicology

Continuation of Residue Reviews

Editor
George W. Ware

VOLUME 163

Springer

Coordinating Board of Editors

DR. GEORGE W. WARE, *Editor*
Reviews of Environmental Contamination and Toxicology

5794 E. Camino del Celador
Tucson, Arizona 85750, USA
(520) 299-3735 (phone and FAX)

DR. HERBERT N. NIGG, *Editor*
Bulletin of Environmental Contamination and Toxicology

University of Florida
700 Experimental Station Road
Lake Alfred, Florida 33850, USA
(941) 956-1151; FAX (941) 956-4631

DR. DANIEL R. DOERGE, *Editor*
Archives of Environmental Contamination and Toxicology

6022 Southwind Drive
N. Little Rock, Arkansas, 72118, USA
(501) 791-3555; FAX (501) 791-2499

Springer-Verlag
New York: 175 Fifth Avenue, New York, NY 10010, USA
Heidelberg: Postfach 10 52 80, 69042 Heidelberg, Germany

Printed in the United States of America.
ISBN 978-1-4419-3188-7
ISSN 0179-5953

Printed on acid-free paper.

Foreword

International concern in scientific, industrial, and governmental communities over traces of xenobiotics in foods and in both abiotic and biotic environments has justified the present triumvirate of specialized publications in this field: comprehensive reviews, rapidly published research papers and progress reports, and archival documentations. These three international publications are integrated and scheduled to provide the coherency essential for nonduplicative and current progress in a field as dynamic and complex as environmental contamination and toxicology. This series is reserved exclusively for the diversified literature on "toxic" chemicals in our food, our feeds, our homes, recreational and working surroundings, our domestic animals, our wildlife and ourselves. Tremendous efforts worldwide have been mobilized to evaluate the nature, presence, magnitude, fate, and toxicology of the chemicals loosed upon the earth. Among the sequelae of this broad new emphasis is an undeniable need for an articulated set of authoritative publications, where one can find the latest important world literature produced by these emerging areas of science together with documentation of pertinent ancillary legislation.

Research directors and legislative or administrative advisers do not have the time to scan the escalating number of technical publications that may contain articles important to current responsibility. Rather, these individuals need the background provided by detailed reviews and the assurance that the latest information is made available to them, all with minimal literature searching. Similarly, the scientist assigned or attracted to a new problem is required to glean all literature pertinent to the task, to publish new developments or important new experimental details quickly, to inform others of findings that might alter their own efforts, and eventually to publish all his/her supporting data and conclusions for archival purposes.

In the fields of environmental contamination and toxicology, the sum of these concerns and responsibilities is decisively addressed by the uniform, encompassing, and timely publication format of the Springer-Verlag (Heidelberg and New York) triumvirate:

Reviews of Environmental Contamination and Toxicology [Vol. 1 through 97 (1962–1986) as Residue Reviews] for detailed review articles concerned with any aspects of chemical contaminants, including pesticides, in the total environment with toxicological considerations and consequences.

Bulletin of Environmental Contamination and Toxicology (Vol. 1 in 1966) for rapid publication of short reports of significant advances and discoveries in the fields of air, soil, water, and food contamination and pollution as well as

methodology and other disciplines concerned with the introduction, presence, and effects of toxicants in the total environment.

Archives of Environmental Contamination and Toxicology (Vol. 1 in 1973) for
important complete articles emphasizing and describing original experimental or theoretical research work pertaining to the scientific aspects of chemical contaminants in the environment.

Manuscripts for *Reviews* and the *Archives* are in identical formats and are
peer reviewed by scientists in the field for adequacy and value; manuscripts for
the *Bulletin* are also reviewed, but are published by photo-offset from camera-
ready copy to provide the latest results with minimum delay. The individual editors of these three publications comprise the joint Coordinating Board of Editors
with referral within the Board of manuscripts submitted to one publication but
deemed by major emphasis or length more suitable for one of the others.

                                                        Coordinating Board of Editors

Preface

Thanks to our news media, today's lay person may be familiar with such environmental topics as ozone depletion, global warming, greenhouse effect, nuclear and toxic waste disposal, massive marine oil spills, acid rain resulting from atmospheric SO_2 and NO_x, contamination of the marine commons, deforestation, radioactive leaks from nuclear power generators, free chlorine and CFC (chlorofluorocarbon) effects on the ozone layer, mad cow disease, pesticide residues in foods, green chemistry or green technology, volatile organic compounds (VOCs), hormone- or endocrine-disrupting chemicals, declining sperm counts, and immune system suppression by pesticides, just to cite a few. Some of the more current, and perhaps less familiar, additions include *xenobiotic transport, solute transport, Tiers 1 and 2, USEPA to cabinet status, and zero-discharge*. These are only the most prevalent topics of national interest. In more localized settings, residents are faced with leaking underground fuel tanks, movement of nitrates and industrial solvents into groundwater, air pollution and "stay-indoors" alerts in our major cities, radon seepage into homes, poor indoor air quality, chemical spills from overturned railroad tank cars, suspected health effects from living near high-voltage transmission lines, and food contamination by "flesh-eating" bacteria and other fungal or bacterial toxins.

It should then come as no surprise that the '90s generation is the first of mankind to have become afflicted with *chemophobia*, the pervasive and acute fear of chemicals.

There is abundant evidence, however, that virtually all organic chemicals are degraded or dissipated in our not-so-fragile environment, despite efforts by environmental ethicists and the media to persuade us otherwise. However, for most scientists involved in environmental contaminant reduction, there is indeed room for improvement in all spheres.

Environmentalism is the newest global political force, resulting in the emergence of multi-national consortia to control pollution and the evolution of the environmental ethic. Will the new politics of the 21st century be a consortium of technologists and environmentalists or a progressive confrontation? These matters are of genuine concern to governmental agencies and legislative bodies around the world, for many serious chemical incidents have resulted from accidents and improper use.

For those who make the decisions about how our planet is managed, there is an ongoing need for continual surveillance and intelligent controls to avoid endangering the environment, the public health, and wildlife. Ensuring safety-in-use of the many chemicals involved in our highly industrialized culture is a dynamic challenge, for the old, established materials are continually being displaced by newly developed molecules more acceptable to federal and state regulatory agencies, public health officials, and environmentalists.

Adequate safety-in-use evaluations of all chemicals persistent in our air, foodstuffs, and drinking water are not simple matters, and they incorporate the judgments of many individuals highly trained in a variety of complex biological, chemical, food technological, medical, pharmacological, and toxicological disciplines.

Reviews of Environmental Contamination and Toxicology continues to serve as an integrating factor both in focusing attention on those matters requiring further study and in collating for variously trained readers current knowledge in specific important areas involved with chemical contaminants in the total environment. Previous volumes of *Reviews* illustrate these objectives.

Because manuscripts are published in the order in which they are received in final form, it may seem that some important aspects of analytical chemistry, bioaccumulation, biochemistry, human and animal medicine, legislation, pharmacology, physiology, regulation, and toxicology have been neglected at times. However, these apparent omissions are recognized, and pertinent manuscripts are in preparation. The field is so very large and the interests in it are so varied that the Editor and the Editorial Board earnestly solicit authors and suggestions of underrepresented topics to make this international book series yet more useful and worthwhile.

Reviews of Environmental Contamination and Toxicology attempts to provide concise, critical reviews of timely advances, philosophy, and significant areas of accomplished or needed endeavor in the total field of xenobiotics in any segment of the environment, as well as toxicological implications. These reviews can be either general or specific, but properly they may lie in the domains of analytical chemistry and its methodology, biochemistry, human and animal medicine, legislation, pharmacology, physiology, regulation, and toxicology. Certain affairs in food technology concerned specifically with pesticide and other food-additive problems are also appropriate subjects.

Justification for the preparation of any review for this book series is that it deals with some aspect of the many real problems arising from the presence of any foreign chemical in our surroundings. Thus, manuscripts may encompass case studies from any country. Added plant or animal pest-control chemicals or their metabolites that may persist into food and animal feeds are within this scope. Food additives (substances deliberately added to foods for flavor, odor, appearance, and preservation, as well as those inadvertently added during manufacture, packing, distribution, and storage) are also considered suitable review material. Additionally, chemical contamination in any manner of air, water, soil, or plant or animal life is within these objectives and their purview.

Normally, manuscripts are contributed by invitation, but suggested topics are welcome. Preliminary communication with the Editor is recommended before volunteered review manuscripts are submitted.

Tucson, Arizona   G.W.W.

Table of Contents

Rev Environ Contam Toxicol 163:1–28

Application of Chitosan for Treatment of Wastewaters

Hong Kyoon No and Samuel P. Meyers

Contents

I. Introduction

Significant volumes of wastewaters, with organic and inorganic contaminants such as suspended solids, dyes, pesticides, toxicants, and heavy metals, are discharged from various industries. These wastewaters create a serious environmental problem and pose a threat to water quality when discharged into rivers and lakes. Thus, such contaminants must be effectively removed to meet increasingly stringent environmental quality standards. It is becoming increasingly recognized that the nontoxic and biodegradable biopolymer chitosan can be used in wastewater treatment (Peniston and Johnson 1970).

Chitosan is a modified, natural carbohydrate polymer derived by deacetylation of chitin, a major component of the shells of crustacea such as crab, shrimp, and crawfish and the second most abundant natural biopolymer after cellulose (No and Meyers 1989a) (Fig. 1). During the past several years, chitosan has received increased attention for its commercial applications in the biomedical, food, and chemical industries (Knorr 1984; Muzzarelli 1973; Sandford and Hutchings 1987). Its major applications currently are in industrial wastewater treatment. Chitosan with high free amino groups can effectively function as a

Communicated by Douglas L. Park.

H. K. No (✉)
Department of Food Science and Technology, Catholic University of Taegu-Hyosung, Hayang 712-702, South Korea
S. P. Meyers
Department of Food Science, Louisiana State University, Baton Rouge, LA 70803, U.S.A.

coagulant and as an adsorbent in wastewater treatment (Peniston and Johnson 1970).

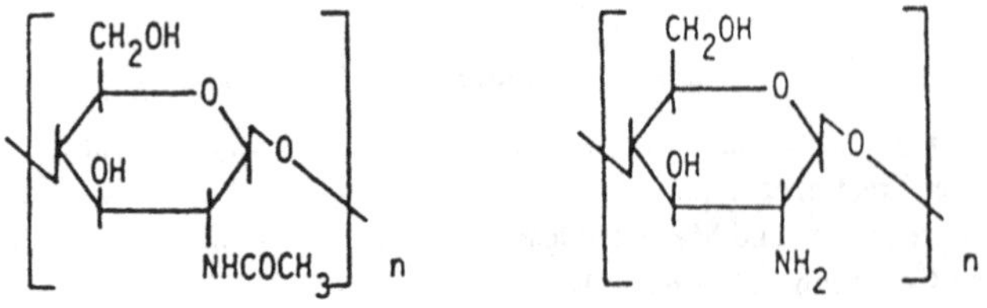

Fig. 1. Chemical structures of chitin and chitosan.

Earlier investigations have demonstrated the effectiveness of chitosan for coagulation and recovery of suspended solids in food processing wastes, with reduction in suspended solids of 65%–99% (Bough 1975b; Jun et al. 1994; Senstad and Almas 1986). Chitosan has also been applied as a chelating polymer for binding harmful metal ions, such as copper, lead, mercury, and uranium, from wastewater (Masri et al. 1974; Muzzarelli 1973). The effectiveness of chitosan for its ability to chelate transition metal ions has been reported by numerous workers (Masri and Randall 1978; Muzzarelli 1973; Yang and Zall 1984). In other areas, chitosan has been employed as an excellent adsorbent for sorption of dyes (Michelsen et al. 1993; Smith et al. 1993), phenols (Sun et al. 1992; Wada et al. 1993), and polychlorined biphenyls (PCBs) (Thomé and Van Daele 1986; Van Daele and Thomé 1986) from wastewater.

This review focuses on the application of chitosan for treatment of wastewaters containing suspended solids, heavy metals, dyes, phenols, and PCBs.

II. Wastewater Treatment

A. Food Processing and Manure Wastewater

Food Processing Wastewater. The major commercial applications for chitosan currently are found in industrial wastewater treatment and in recovery of feed-grade material from food processing plants. Chitosan carries a partial positive charge; thus, it functions effectively as a polycationic coagulating agent in wastewater treatment applications (Peniston and Johnson 1970).

A mechanism of action of polymeric flocculating agents is described by LaMer and Healy (1963) in which the polymer destabilizes a colloidal suspension by adsorption of particles with subsequent formation of particle–polymer–particle bridges. This is generally true both for anionic and nonionic polyelectrolytes that are used to coagulate negative colloid. Positively charged cationic polymers can destabilize a negative colloidal suspension by charge neutralization as well as by bridge formation (O'Melia 1972).

Various studies have demonstrated the effectiveness of chitosan for coagulation and recovery of suspended solids in processing wastes from a variety of industries including poultry (Bough et al. 1975), eggs (Bough 1975a), cheese (Bough and Landes 1976; Wu et al. 1978), meat and fruit cakes (Bough 1976), seafood (Bough 1976; Johnson and Gallanger 1984; No and Meyers 1989a; Senstad and Almas 1986), vegetable operations (Bough 1975b; Moore et al. 1987), and tofu (Jun et al. 1994). These studies indicate that chitosan can reduce the suspended solids of such food processing wastes by as much as 65% to 99%. Table 1 summarizes data from various studies on reduction of turbidity (TB), suspended solids (SS), and chemical oxygen demand (COD) in food processing waste effluents by coagulation with chitosan followed by gravity settling. In addition to reduction of waste load, the coagulated by-products recovered from food processing wastes with chitosan contain significant amounts (13%–72%) of protein (Table 2) and offer potential sources of protein in animal feeds (Bough and Landes 1978).

Table 1. Reduction of turbidity (TB), suspended solids (SS), and chemical oxygen demand (COD) in food processing waste effluents by coagulation with chitosan and gravity settling (GS).

Effluent	Chitosan (mg/L)	GS time (hr)	Percentage of reduction			Reference
			TB	SS	COD	
Egg-breaking composite	150[a]	1.5	78	72	76	1
	200[b]	1.5	77	74	57	1
Egg washer waste	100[c]	1.5	95	94	60	1
Greens washer	10[d]	3	90	99	62	2
Greens filler	5[e]	1.5	83	93	10	2
Greens composite	10	1	85	90	—	2
Spinach composite	20	1	94	90	—	2
Pimento peeling	40	1	82	96	—	2
Pimento coring	10	1	64	84	—	2
Pimento composite	20	1	77	89	—	2
Green bean blancher	5[f]	1	91	95	—	2
Meat packing	30	1.5	96	89	55	3
Meat processing and curing	10[g]	1.5	84	95	72	3
	5	—	—	92	79	3
Shrimp processing	10[h]	1	97	94	76	3
	60–360	1	85	65[i]	—	8
	30	0.5	98	98	—	9

Table 1. (Continued).

Effluent	Chitosan (mg/L)	GS time (hr)	Percentage of reduction			Reference
			TB	SS	COD	
Crab processing	30	0.5	96	98	—	9
Salmon processing	30	0.5	97	98	—	9
Fruit cake processing	2	2	89	94	47	3
Cheese whey	53	3	—	92	4	4
Poultry composite	5	1	93	74–94	13	5
Poultry chiller	6	1	75	75	62	5
Poultry scalder	30	1.5	87	88	49	5
Crawfish processing	150	1	83	97	45	6
Tofu	300	1	97	—	—	7

[a]Plus 10 mg/L Betz 1130.
[b]Plus 15 mg/L Betz 1130.
[c]Plus 20 mg/L Betz 1130.
[d]Plus 15 mg/L NJAL-240
[e]Plus 10 mg/L NJAL-240 and 40 mg/L alum.
[f]Plus 80 mg/L $CaCl_2$.
[g]Plus 40 mg/L $FeCl_3$.
[h]Plus 5 mg/L WT-3000.
[i]Percentage of protein removal.
References: 1, Bough (1975a); 2, Bough (1975b); 3, Bough (1976); 4, Bough and Landes (1976); 5, Bough et al. (1975); 6, No and Meyers (1989a); 7, Jun et al. (1994); 8, Senstad and Almas (1986); 9, Johnson and Gallanger (1984).

Table 2. Proximate composition of coagulated solids recovered from food processing waste effluents by coagulation with chitosan and gravity settling.

Effluent	Solids composition (%)			Reference
	Protein	Fat	Ash	
Egg-breaking composite	42.2–42.9[a]	40.8–45.3	2.9–8.8	1
Egg washer waste	49.4[a]	36.1	4.2	1
Meat packing	41[a]	17	11	3
Fruit cake	13[a]	—[d]	—	3
Cheese whey	72.3[b]	0.2	9.5	4
Poultry composite	54.0[a]	29.4	4.3	5
Poultry chiller	34.4[a]	57.7	1.3[e]	5

Table 2. (Continued).

Effluent	Solids composition (%)			Reference
	Protein	Fat	Ash	
Poultry scalder	70.6[a]	1.2	15.7	5
Crawfish processing	27.1[a]	51.7	3.3	6
Tofu	41.9[c]	1.5	0.9	7

[a]Based on Kjeldahl N ×6.25.
[b]Based on Kjeldahl N ×6.4.
[c]Based on total amino acid content.
[d]Not determined.
[e]Ash value of solids recovered by dissolved air flotation.
References: see Table 1.

Bough et al. (1975) observed that treatment of poultry processing wastes with 5 mg/L chitosan as a coagulating agent reduced suspended solids in the composite effluents by 74%–94%. The dry coagulated solids contained 54.0% crude protein, 29.4% fat, and 4.3% ash. Proteinaceous solids recovered from cheese whey, with and without chitosan as a coagulating agent, had protein efficiency ratios equivalent to that of the casein control (Bough and Landes 1976).

In some instances, chitosan has been used in conjunction with a synthetic polyelectrolyte or an inorganic salt to increase treatment effectiveness. With egg-breaking wastes (Bough 1975a), 100–200 mg/L chitosan and 10–20 mg/L of the synthetic coagulant Betz 1130 effectively coagulated suspended solids in wastewater. These solids were reduced by 72%–94% and COD by 57%–76%. The amino acid composition of coagulated egg by-products was comparable to that of whole eggs.

In general, suspended solids in food processing wastes are often reduced by 90% or more with coagulation. On the other hand, COD reductions are usually much less because of the high content of dissolved organic substances that are not removed by coagulation (Bough et al. 1975). However, in certain instances, such as with poultry (Bough et al. 1975), eggs (Bough 1975a), and meat and seafood wastes (Bough 1976), COD reductions of 60%–80% have been obtained.

No and Meyers (1989b) investigated the recovery of amino acids from crawfish processing wastewater using amino copper–chitosan. The amino acids recovered by this treatment had potential application as seafood flavors based on their sensory attributes.

Chitosan also is effective for conditioning activated sludge in conjunction with centrifugal dewatering (Bough 1976). Treatment of activated brewery and vegetable sludge with 75 and 40 mg/L chitosan resulted in removal of suspended solids by 95% and 99%, respectively. Asano et al. (1978) found that chitosan is effective for dewatering municipal and industrial sludges. Among various polyelectrolytes, the chitin–chitosan-derived polymers known as Flonac in Japan have been widely used for sludge dewatering applications, mainly because of their effectiveness in sludge conditioning, rapid biodegradability in soil environments and economic advantages in centrifugal sludge dewatering (Asano et al. 1978).

Manipulation in the chitosan manufacturing process produces chitosans with varying chemical characteristics and molecular weight distributions (Wu and Bough 1978). These products differ in their effectiveness as waste treatment agents for conditioning of activated sludge and coagulation of food processing wastewater. For example, Wu and Bough (1978) found that high-viscosity chitosan products that were more effective for vegetable sludge conditioning were less effective for coagulation of cheese whey. These workers postulated that this was primarily caused by the difference in particle size or charge characteristics of the two wastewater systems, suggesting that different chitosan samples can be produced for particular wastewater systems. Bough et al. (1978) compared various chitosan samples prepared under different conditions for effectiveness of coagulation of an activated vegetable sludge suspension. The results indicated that treatment effectiveness did not correlate linearly with viscosity and molecular weight distribution of products; however, higher values for molecular weight were predictive of greater effectiveness for coagulation of activated sludge suspensions.

Manure Wastewater. There is little information on treatment of wastewaters from livestock and poultry enterprises using chitosan. Sievers et al. (1994) recently studied chemical coagulation of suspended solids from dilute manure wastewaters [fresh manure (SN) from swine, poultry, and cattle, and effluent from a swine anaerobic digester (DE)] using various coagulants ($FeCl_3$, chitosan, and synthetic polymers). Excellent removals (75%) of volatile solids were obtained with chitosan when applied to cattle and swine SN at 100–150 mg/L chitosan. A maximum of 50% volatile solids reduction was achieved with poultry SN at 150 mg/L chitosan and even less removal was obtained with swine DE. The pH of swine and cattle SN solutions after addition of the chitosan ranged from 6.0 to 6.8. Poultry SN and DE were alkaline solutions (pH 7.5–8.5) because of high ammonia levels. The high pH of these wastewaters most likely accounts for the poor performance of chitosan as a coagulant because chitosan is insoluble in alkaline solutions and will precipitate at a solution pH greater than 6.5–7.0.

In these studies, ferric chloride removed 60%–70% of the volatile solids for cattle and swine SN, and about 36% for poultry SN at concentrations of 300–450 mg/L. Synthetic polymers 767 (cationic) and 725 (anionic) removed the volatile solids by 42% and 23%, respectively, from the swine SN at a concentration of 1.0 mg/L polymer. When applied to swine DE, both polymers could reduce turbidity by 85% at 300 mg/L.

B. Metal-Contaminated Wastewater

Wastewater containing a variety of heavy metals is discharged from many sources, such as mining operations, metal plating facilities, fertilizer manufactures, or electronic device manufacturing operations (Rorrer et al. 1993; Yang and

Zall 1984). Because these are often toxic at low concentrations and are not biodegradable, they must be physically removed from the contaminated water to meet increasingly stringent environmental quality standards (Rorrer et al. 1993). Adsorption techniques, specifically those using chelating resin, have been widely demonstrated and promoted as being feasible technology (Deans and Dixon 1992; Yang and Zall 1984).

Chitosan is very attractive for heavy metal ion separations from wastewater because it selectively binds to virtually all group III transition metal ions at low concentrations but does not bind to groups I and II alkali and alkaline earth metal ions (Muzzarelli 1973, 1997). The amino groups on the chitosan chain serve as a chelation site for transition metal ions (Rorrer et al. 1993).

Muzzarelli (1973) documented the effectiveness of chitin, chitosan, and other polymers for chelating of transition metal ions. He indicated that chitosan was a powerful chelating agent and exhibited the best collection ability of all the tested polymers because of its high amino group. Masri et al. (1974) compared the chelating ability of chitosan with that of several other materials, i.e., bark, activated sewage sludge, and poly(p-aminostyrene), and confirmed that chitosan had a higher chelation ability. Yang and Zall (1984) found that chitin, chitosan, and scales from three species of fish (porgy, flounder, and cod) were potentially useful materials to remove metals [Cu, Zn, Cr(III), Cd, and Pb] from contaminated water; however, the diffusion of solutes through internal particles of chitosan was relatively faster than the other adsorbents tested.

Jha et al. (1988) studied the uptake of cadmium with time at pH 6.5 at initial concentrations of 1.5, 5.0, and 10.0 mg/L and found that the rate of sorption was very rapid initially and decreased markedly after 4 hr. According to these workers, amino groups of chitosan act initially as cadmium coordination sites, and the slower removal rate of metal after an initial rapid uptake can be caused by the binding of cadmium by complexed metal ions. The relatively rapid initial rate of adsorption was observed with mercuric ions by Peniche-Covas et al. (1992).

There are essentially three consecutive stages associated with the adsorption of materials from solution by porous adsorbents: transport of the adsorbate to the exterior surface of the adsorbent, diffusion of the adsorbate into the pores of the adsorbent, and adsorption of the solute on the interior surfaces of adsorbent. In general, the adsorption of solute onto the surface of an adsorbent is relatively fast compared to the other two processes. In addition, external resistance is small if sufficient mixing is provided (Yang and Zall 1984). The study of adsorption kinetics of copper by chitosan showed that the intraparticle diffusion was the rate-determining step (Yang and Zall 1984). This mechanism was observed subsequently by McKay et al. (1986) and Peniche-Covas et al. (1992) with Cu and Hg ions, respectively.

The free amino group in chitosan was considered to be much more effective for binding metal ions than the acetyl groups in chitin (Eiden et al. 1980; Maruca et al. 1982; Muzzarelli and Tubertini 1969; Suder and Wightman 1983; Yang and Zall 1984). This information leads us to consider that the greater free amino group content of chitosan should give higher metal ion adsorption rates. Yang and

Zall (1984) noted that chitosan has many free amino groups, chelating as much as five- to sixfold greater amounts of metals than does chitin. Similarly, Eiden et al. (1980) reported that the uptake of Pb(II) on chitin was approximately 21% of that on chitosan. However, Deans and Dixon (1992) studied comparative efficiencies of a series of different functionalized biopolymers for removing Pb(II) and Cu(II) ions from water at 1, 10, and 100 ppm and observed that there is no optimal biopolymer for metal ion uptake. The uptake results varied with the metal ion identity and its concentration and with adsorbent structure. For example, for Cu and Pb at 1 ppm, the best adsorbent was found to be chitosan and chitin, respectively. For Pb at 10 and 100 ppm, the carboxymethylcellulose hydroxamic acid was the best performer. These investigators concluded that the choice of an ideal adsorbent should be determined for each application.

The adsorption ability of chitosan is dependent on many other factors, such as crystallinity, deacetylation, and hydrophilicity (Kurita et al. 1979). Earlier, Muzzarelli and Tubertini (1969) also mentioned that adsorption depends on time of contact, temperature, pH, concentration of the ion under examination, and concentration of the other ions present.

It is interesting to note the relationship between physicochemical properties or sources of chitosan and metal-binding property. Madhavan and Nair (1978) studied metal-binding (Zn, Ni, Cr, Cu, Fe, Mn) properties of three chitosans with viscosities of 155, 430, and 902 cP, respectively, in a 1% solution in 1% acetic acid. It was found that the metal-binding property of chitosan was not affected by viscosity characteristics. Nair and Madhavan (1982) investigated metal (Fe, Co, Ni, Hg, Cu) binding property of chitosans prepared from four different sources of crab (*Scylla serrata*), prawn (*Penaeus indicus*), squid (*Loligo* sp.), and squilla (*Oratosquilla nepa*). These four chitosans studied did not show notable variations in the adsorption rate (Table 3). Kurita et al. (1979) found that a homogeneous hydrolysis process could give a chitosan product with a higher adsorption rate for metal ions than one prepared by a heterogeneous process with the same degree of deacetylation.

Particle size may also affect the metal adsorption capacity of chitosan. Decreasing particle size increased the adsorption capacity for Cd (II) (Jha et al. 1988; Rorrer et al. 1993), Cr(III) (Maruca et al. 1982), and Cr(VI) (Udaybhaskar et al. 1990). On the other hand, McKay et al. (1986) found only a small variation in Cu(II) adsorption capacity per unit mass of chitosan with particle size (0.21–1.0 mm). The increase in adsorption capacity with decreasing particle size suggests that the metal preferentially adsorbed on the outer surface and did not fully penetrate the particle. The formation of adsorbed metal clusters is supported by Eiden et al. (1980) and Maruca et al. (1982), who observed Cr-containing nodules on the surface of Cr(III)-equilibrated chitosan powder using SEM/EDAX (scanning electron microscope/energy dispersive analysis of X-rays) analysis. Maruca et al. (1982) and Suder and Wightman (1983) further investigated the uptake mechanism of Cr(III), Cd(II), and Zn(II) by chitin and chitosan using SEM/EDAX analysis, and concluded that a combination of nodular formation, ion adsorption, and ion absorption accounts for the total uptake.

Table 3. Rate of adsorption of metal ions by chitosan from different sources.

Metal ions	Chitosan source	mg of metal adsorbed/g of chitosan			
		10 min[a]	30 min	60 min	120 min
Fe^{3+}	Crab	11.7	17.6	23.4	23.4
	Prawn	5.9	11.7	15.7	23.4
	Squid	17.6	17.6	20.5	23.4
	Squilla	11.7	14.6	17.6	29.3
Co^{2+}	Crab	0.0	4.1	4.7	5.9
	Prawn	4.7	5.3	7.1	7.1
	Squid	4.7	4.7	4.7	7.4
	Squilla	4.7	4.7	4.7	4.7
Ni^{2+}	Crab	11.7	35.2	55.2	64.6
	Prawn	29.3	47.0	64.6	82.1
	Squid	11.7	29.3	64.6	82.1
	Squilla	5.8	29.3	52.8	76.3
Hg^{2+}	Crab	161	241	281	321
	Prawn	251	311	331	341
	Squid	200	321	346	366
	Squilla	261	351	381	411
Cu^{2+}	Crab	15.1	21.1	36.2	39.3
	Prawn	27.2	30.2	42.3	66.4
	Squid	12.1	27.2	45.3	51.3
	Squilla	9.0	42.3	60.4	60.4

[a]Shaking time.
Source: Nair and Madhavan (1982).

Differences in pH of the solution have been reported to influence the metal ion adsorption capacity of chitosan. The uptake of Cd(II) (Jha et al. 1988; Suder and Wightman 1983) and Cr(III) (Maruca et al. 1982) ions by chitosan decreased with decreasing pH. These results suggest that hydronium ions compete with metal ions for available amino sites on chitosan (Rorrer et al. 1993). However, drastic decrease in Cr (VI) removal was observed with an increase in pH to neutral and above by Udaybhaskar et al. (1990). Sorption was almost 90% at pH 3 at an initial Cr concentration of 5 mg/L and was reduced to 10% at pH 7 and above. In contrast, Nair and Madhavan (1982) found that pH (2.9–8.4 for Co^{2+}, 2.9–6.5 for Ni^{2+}, and 3.0–6.0 for Hg^{2+}) of the solution did not influence significantly either the rate of adsorption or the quantity of metal adsorbed by chitosan.

McKay et al. (1986, 1989) measured adsorption isotherms for Cu^{2+}, Hg^{2+}, Ni^{2+}, and Zn^{2+} on chitosan powder as a function of temperature (25–60°C) and found that adsorption capacity decreased with increasing temperature. The presence of counterions or other substances in solution affects the adsorption of metals by chitosan. The presence of the sulfate ion enhanced the adsorption of Cu and Ni ions by swollen chitosan beads (Mitani et al. 1991), and the uptake of Cr(III) by chitosan was enhanced in the presence of 0.05 M phosphate at pH 5.0 (Maruca et al. 1982). Jha et al. (1988) reported that the presence of appreciable quantities of chloride did not have any effect on cadmium removal. Furthermore, removal of cadmium by chitosan was not significantly affected by the presence of calcium ions at 100 ppm, but was significantly decreased by zinc or complexing agents such as EDTA.

There is a great variation in the affinity of chitosan for different metals in a single solute system (Ishii et al. 1995; Madhavan and Nair 1978; Nair and Madhavan 1982), as well as in a multisolute system (Yang and Zall 1984). In a single solute system, for example, the order of extent of removal (%) by chitosan at metal ion concentrations of 100 ppm was Cu (98.3) > Ni (78.5) > Co (21.0) > Mn (7.0) (Ishii et al. 1995). Nair and Madhavan (1982) found significant variation in affinity of the different metal ions (Fe, Co, Ni, Hg, Cu) to chitosan, with minimum adsorption for Co^{2+} (7.4 mg/g) and maximum for Hg^{2+} (411 mg/g). In a multisolute system, Yang and Zall (1984) found that the selectivity sequence in uptake of metals by chitosan was Cu > Cr = Cd > Pb > Zn. The percentage removal of metals by chitosan was shown to be approximately 10%–15% less in a multisolute system than in a single solute system because competitive adsorption occurs in a multisolute system. However, Ishii et al. (1995) noted that metal ions can be removed almost completely under competitive conditions using amino acid ester-substituted chitosans.

Jha et al. (1988) demonstrated that 88% of the cadmium adsorbed on chitosan powder could be released in 0.01 N HCl. Under low-pH conditions, protons effectively compete for the active sites, thus releasing the metal ion into the suspending medium. Coughlin et al. (1990) also showed that Ni, Cu, and Cr(III) adsorbed on chitosan flake could be desorbed at low pH, even under repeated adsorption–desorption cycles. Thus, the metal ion adsorption process on chitosan is reversible, making adsorbent regeneration and metal recovery at low pH feasible (Rorrer et al. 1993). On the other hand, Masri and Randall (1978) indicated that the copper bound to chitosan was readily desorbed with a solution of ammonium hydroxide-ammonium chloride of pH 10. Minimal desorption of chromium was observed on washing previously chromium-equilibrated chitosan with distilled water (Maruca et al. 1982).

Masri and Randall (1978) demonstrated the effectiveness of chitosan for removing toxic metal ions in the actual waste streams. Wastes treated were (a) from electroplating and metal-finishing operations (with disposal problems mainly of cyanide and salts of Cr, Cd, Zn, Pb, Cu, Fe, and Ni); (b) from a nickel–salt manufacturing plant (disposal of nickel and alkali); (c) from a lead-battery manufacturing plant (disposal of sulfuric acid and lead salts); and (d) from exhausted dyebath for wool fabrics in which dichromate is included in the bath (disposal of chromium). Chitosan was effective in reducing the content of copper, cadmium, iron, zinc, lead and nickel salts, and sulfuric acid. For example, treatment of nickel waste with chitosan (500 mL/2 g for 1 hr) reduced the nickel ion content from 7.2 to 1.2 ppm. Treatment of lead-battery wastes with chitosan (50 mL/2 g for 1 hr) reduced the lead concentration from 1.8 to less than 0.5 ppm and the iron concentration from 30 to 8.7 ppm.

Coughlin et al. (1990) studied the potential of using chitosan for purifying electroplating wastewater. Partially deacetylated chitin (PDC, deacetylated at the outer periphery of the particles without substantially deacetylating the interior regions of the particles) was used as a metal ion-complexing material. PDC was found to be a useful sorbent for transition metal ions (Cu, Ni, Cr), being nearly as effective as relatively pure chitosan. Thus, Coughlin et al. (1990) concluded this approach to be economically more favorable than the conventional precipitation process under reasonable assumptions.

Other workers (Masri and Randall 1978; Yang and Zall 1984) have suggested that although chitosan as such is useful for treating wastewaters, it may be advantageous to use the cross-linked chitosan with di- or polyfunctional reagents to impart increased resistance to solubilization in acidic pH effluents. Chitosan is insoluble in aqueous alkali solution of pH up to 13 or in inorganic solvents, but is soluble in weak acid solutions, except for sulfuric acid (Hauer 1978). This solubility undoubtedly limits its use as an adsorbent in low-pH regions.

More recent studies showed that the chelating ability of chitosan could be further improved by chemical modifications such as cross-linking with glutaraldehyde (Koyama and Taniguchi 1986; Kurita 1987) (Fig. 2A), acetylation (Hsien and Rorrer 1995; Kurita 1987; Kurita et al. 1988) (Fig. 2B), and amino acid conjugates (Ishii et al. 1995) (Fig. 3). For example, Koyama and Taniguchi (1986) reported that homogeneously cross-linked chitosan with glutaraldehyde at the aldehyde–amino group ratio of 0.7 increased adsorption of Cu ion as high as 96% in an aqueous solution of pH 5.2, although the original chitosan collected only 74% of Cu ion. This increase in adsorption was interpreted in terms of the increases in hydrophilicity and accessibility of chelating groups as a result of partial destruction of the crystalline structure of chitosan by cross-linking under homogeneous conditions. The X-ray diffraction diagrams revealed a marked reduction in crystallinity by cross-linking, and the cross-linked chitosan was found to be almost amorphous.

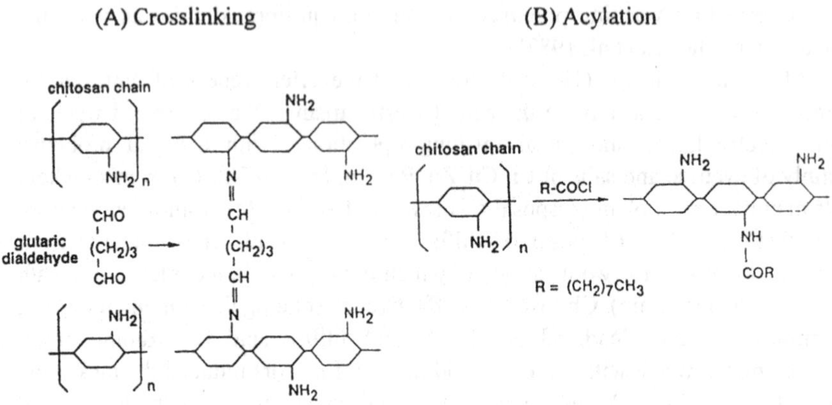

(A) Crosslinking   (B) Acylation

Fig. 2. Chemical structures. (A) Cross-linking of chitosan with glutaraldehyde. (B) N-Acylation of chitosan.

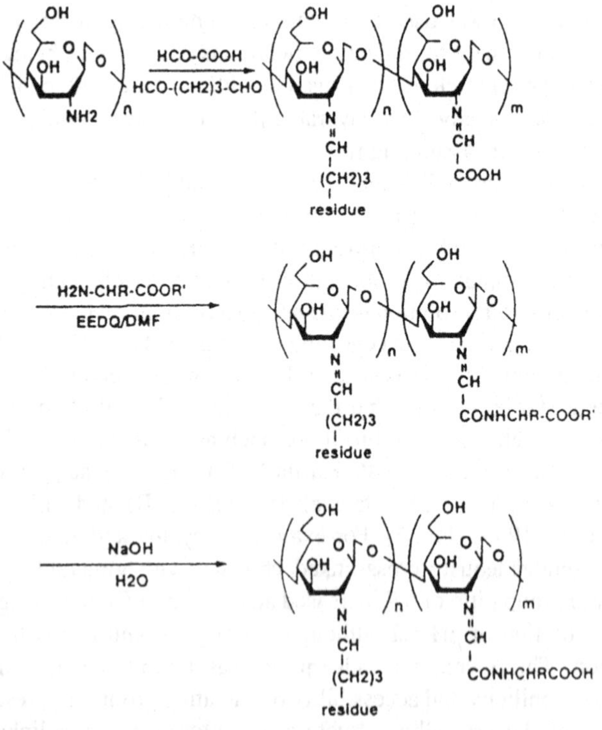

Fig. 3. Synthetic scheme for production of chitosan–amino acid conjugates. R, side chain of amino acid; R′, alcohol component of ester. (From Ishii et al. 1995.)

Masri and Randall (1978), on the other hand, attempted the cross-linking of chitosan with glutaraldehyde under heterogeneous conditions and found the adsorption was decreased by cross-linking, although cross-linking enabled the use of chitosan in acidic media as the result of insolubilization. Consequently, the controlled cross-linking with an appropriate amount of glutaraldehyde under homogeneous conditions effectively enhanced the adsorption capacity of chitosan, producing high potential adsorbents practically applicable in all pH regions without a dissolution problem (Koyama and Taniguchi 1986).

Ishii et al. (1995) prepared chitosan–amino acid conjugates by coupling amino acid esters to the carboxyl group of glyoxylic acid-substituted chitosan. The removal of heavy metals (Cu, Ni, Co, and Mn) was increased by introduction of amino acids to chitosan. Heavy metals were almost completely removed by chitosan–amino acid conjugates from solutions at 100 ppm. The extent of removal was not affected by the type of amino acid introduced.

Many researchers have clearly demonstrated that chitosan has an intrinsically high affinity and selectivity for transition metal ions. However, the adsorbent raw material is not suitable for processing aqueous waste streams. Chitosan is usually obtained in a flaked or powdered form that is both nonporous and soluble in acidic media. The low internal surface area of the nonporous material limits access to interior adsorption sites and hence lowers metal ion adsorption capacities and adsorption rates. The solubility of chitosan in acidic media prevents its use in recovery of metal ions from wastewater at low pH. Furthermore, the flaked or powdered form of chitosan swells and crumbles easily and does not function ideally in packed-column configurations common to pump-and-treat adsorption processes (Rorrer et al. 1993).

Recently, porous beads of chitosan were synthesized for adsorption of ppm-level transition metal ions from aqueous solution (Hsien and Rorrer 1995; Kawamura et al. 1993; Rorrer et al. 1993). Rorrer et al. (1993) described the synthesis of highly porous, magnetic beads of chemically cross-linked chitosan for removal of cadmium ions from dilute aqueous solution. Highly porous chitosan beads were prepared by dropwise addition of an acidic chitosan solution into a sodium hydroxide solution precipitation bath. The gelled chitosan beads were cross-linked with glutaraldehyde and then freeze-dried. Beads of 1- and 3-mm diameter possessed internal surface areas of 156 and 94 m^2/g and mean pore sizes of 560 and 540 Å, respectively, and were insoluble in acid media at pH 2. The surface area of the nonporous chitosan flake was only 0.35 m^2/g.

Kawamura et al. (1993) synthesized polyaminated highly porous chitosan chelating resin and found that polyamination of the amine group significantly improved adsorption capacity for mercury ions. Hsien and Rorrer (1995) synthesized N-acylated, cross-linked highly porous chitosan beads insoluble in 1 M acetic acid solution (pH 2.4) with an internal surface area of 224 m^2/g.

The behavior of the adsorption isotherms investigated by Rorrer et al. (1993) suggested that both the size and the porous structure of the bead have a profound effect on adsorption capacity. The internal surface area and porosity of the 1-mm beads were greater than the 3-mm beads, and maximum adsorption capacities for the 1- and 3-mm beads were 518 and 188 mg of Cd/g of bead, respectively. An increase in adsorption capacity with a decrease in particle size also suggests that the metal ions do not completely penetrate the particle and that the metal preferentially adsorbs near the outer surface of the bead by the pore-blockage mechanism.

Table 4 shows literature references studied for adsorption of heavy metals by chitosan and its derivatives.

Table 4. References studied for adsorption of heavy metals by chitosan and its derivatives.

Metal	References
Hg	1, 12, 15, 19, 22, 26, 30, 32
Cd	1, 2, 12, 13, 16, 18, 20, 21, 22, 29, 30, 31
Pb	1, 3, 12, 13, 20, 25, 30
Zn	1, 2, 5, 12, 13, 19, 20, 22, 29, 30, 31, 32
Co	1, 9, 12, 15, 19, 27, 29, 30
Ni	1, 4, 5, 9, 10, 12, 15, 22, 27, 28, 29, 30, 32
Cr	1, 3, 5, 6, 10, 12, 13, 14, 20, 30
Cu	1, 4, 5, 7, 8, 9, 10, 11, 12, 13, 15, 17, 19, 20, 22, 23, 25, 26, 27, 29, 30, 32, 33
Fe	1, 5, 12, 15, 31
Mn	1, 5, 9, 12, 29, 31
Ag	1, 12, 19
Au	1, 12, 19
Pt	1, 12, 24
Pd	1, 19, 24
Mo	1, 19
Sb	19
Ir	19, 24
U	22, 29, 30

References: 1, Masri et al. (1974); 2, Suder and Wightman (1983); 3, Eiden et al. (1980); 4, Mitani et al. (1991); 5, Madhavan and Nair (1978); 6, Maruca et al. (1982); 7, Koyama and Taniguchi (1986); 8, Kurita et al. (1988); 9, Ishii et al. (1995); 10, Coughlin et al. (1990); 11, Peniche-Covas et al. (1992); 12, Masri and Randall (1978); 13, Yang and Zall (1984); 14, Udaybhaskar et al. (1990); 15, Nair and Madhavan (1982); 16, Jha et al. (1988); 17, Kurita (1987); 18, Rorrer et al. (1993); 19, Muzzarelli and Tubertini (1969); 20, Hauer (1978); 21, Hsien and Rorrer (1995); 22, Kawamura et al. (1993); 23, Holme and Hall (1991); 24, Inoue et al. (1995); 25, Deans and Dixon (1992); 26, Ohga et al. (1987); 27, Muzzarelli et al. (1978); 28, Randal et al. (1979); 29, Sakaguchi and Nakajima (1982); 30, Muzzarelli and Tanfani (1982); 31, Seo and Kinemura (1989); 32, McKay et al. (1989); 33, McKay et al. (1986).

C. Dye Wastewater

Textile effluents usually contain very small amounts of dyes; however, these are highly detectable and have become a pollution concern. Several difficulties are encountered in removal of dyes from wastewaters in that they are highly stable molecules, made to resist degradation by light, chemical, biological, and other exposures. Commercial dyes usually are mixtures of large complex with often uncertain molecular structure and properties (Smith et al. 1993).

The adsorption of dyestuffs by solid polymers has been widely investigated, mainly for industrial applications. At least 70 dyes (mainly anionic and direct dyes) have been so far studied in terms of their adsorption on chitin and chitosan (Stefancich et al. 1994). Extensive studies on the adsorption of dyes on chitin by Giles et al. (1958), McKay (1987), and McKay et al. (1982, 1983, 1984, 1985, 1987) have revealed that chitin has the ability to adsorb substantial quantities of dyestuffs from aqueous solutions. However, chitosan with higher amino group content was reported to be more effective for binding dyes than chitin (Park et al. 1995b; Shimizu et al. 1995; Smith et al. 1993). For example, Shimizu et al. (1995) investigated the binding behaviors of chrome violet (C.I. Mordant Violet 5, a monoazo dye) to chitin and partially deacetylated chitin (degree of deacetylation = 65%) and found that the dye-binding capacity of partially deacetylated chitin was very much greater than that of chitin. The content of amino groups in partially deacetylated chitin increased by a factor of 8.6 compared with that in chitin.

The interaction of chitosan with dyes have been studied by several workers. Yamamoto (1984) studied chiral interaction of chitosan with azo dyes (Orange I, Alizarin Yellow GG, and Congo Red) and suggested that intermolecular interactions of the dye molecules are most probable in the chitosan–dye systems. Shimizu et al. (1995) reported that electrostatic interactions are involved in the binding of chrome violet to chitosan. Interaction of dyes with chitosan derivatives was investigated by Seo et al. (1989), who explored the predominant contribution of the hydrophobic interaction in the systems of butyl orange/N-octanoyl chitosan gels.

Recently, the interaction of several cationic and anionic dyes with water-soluble chitosan derivatives (N-carboxymethyl chitosan, N-carboxybutyl chitosan, and the reduction product of the aldimine obtained from chitosan and 5-hydroxymethyl-2-furaldehyde) and the parent chitosan was studied in water (Delben et al. 1994; Stefancich et al. 1994). The occurrence of an interaction in water between the hydrophilic chitosan derivatives and the parent chitosan and the anionic dyes (Orange II, Alizarin S, Alizarin GG, and Congo Red) was proved by optical techniques (Stefancich et al. 1994) and by optical and thermodynamic approaches (Delben et al. 1994). Binding depended on pH, with the pH range 3.5–5 being the most effective. On the contrary, the cationic dyes [Neocuproin, 1-(2-pyridilazo)-2-naphthol, Ethidium bromide, Ruthenium Red) did not interact with the polysaccharides studied in the pH range 5–6 (Stefancich et al. 1994).

Maghami and Roberts (1988) investigated adsorption characteristics of three anionic dyes (C.I. Acid Orange 7, C.I. Acid Red 13, and C.I. Acid Red 27) on chitosan under acid conditions and demonstrated a 1:1 stoichiometry for the interaction of sulfonic acid groups on the dyes with protonated amino groups of the chitosan for mono-, di-, and trisulfonated dyes. Treatment of chitosan with acid produces protonated amine groups along the chain, and these can act as dye sites for anionic dyes. Chitosan, although soluble in dilute acetic acid, will not dissolve if the acetic acid solution contains an excess of an anionic dye. Venkatrao et al. (1986) studied adsorptive behavior of chitosan with respect to two dyes (C.I. Reactive Red 73 and C.I. Direct Red 31) and found that intra-particle diffusion was a major rate-controlling step in the adsorption of direct and reactive dyes from aqueous solutions.

Shimizu et al. (1995) examined the effect of metal ions on the binding of chrome violet by chitin and partially deacetylated chitin. The results indicated that Zn^{2+} and Cu^{2+} ions did not perceptively influence the binding affinity of chrome violet to chitin. In contrast, Co^{2+} ion enhanced binding, and Ni^{2+} ion suppressed it. At lower free dye concentrations, dye uptake by partially deacetylated chitin was greatly enhanced by the addition of Co^{2+} ion in the buffer solution of pH 5. Dye uptake was considerably increased by addition of Cu^{2+} ion at pH 5, becoming much larger at pH 6, the amount corresponding to that in the presence of Co^{2+} ion.

Smith et al. (1993) extensively studied decolorization of dye wastewater using chitosan. In isotherm studies performed on the sorption of Acid Red 1 dye onto chitin, cold-batch chitin, and chitosan, the latter revealed an extremely high sorption capacity for this dye. A study on the effect of flow velocity through a bed of chitosan using a sorption unit demonstrated a complex relationship between flow rate and rate of dye removal from an aqueous dye solution; this probably resulted from complex factors related to internal pore structure, penetration, and diffusion. Nine different dyes also were tested on the prototype decolorization unit using chitosan as an sorbent (Table 5). It was found that the molecular size of the dye was a major factor in sorption characteristics. Small, low molecular weight dyes sorbed best on chitosan while dyes with metal in the structure did not appear to sorb any better than their nonmetallized counterparts.

In equilibrium adsorption measurements with three anionic dyes, Maghami and Roberts (1988) also found that equilibrium was reached more rapidly with the smallest dye, being attained in less than 2 hr with C.I. Acid Orange 7 while approximately 9 hr were required with the other two dyes (C.I. Acid Red 13 and C.I. Acid Red 27). However, the ionic charge on the dye ion appeared to have negligible effect on the time to equilibrium.

Removal of color from Navy 106 slack washwater discharge was conducted using 22 different adsorbents at pH of 10.4 (the pH of the process water), 7.5, and 4.6 by Michelsen et al. (1993). In general, color of the Navy 106 was removed more effectively by sorption with decreased pH. With chitosan, color was removed 36%, 72%, and 86% at pH 10.4, 7.5, and 4.6, respectively. In a study on

Table 5. Sorption characteristics for several dyes on chitosan.

Dye	MW	Chemical type	Original concentration (g/L)	Sorption capacity (mg/g)
Acid Red 1	509	Monoazo	0.0908	7.13
Acid Blue 25	401	Anthraquinone	0.1510	13.39
Acid Blue 193	479	Monoazo	0.1410	8.43
Mordant Black 17	375	Monoazo	0.1120	9.21
Direct Blue 86	781	Phthalocyanine	0.0955	3.94
Direct Red	699	Diazo	0.1299	—
Direct Green	1344	Triazo	0.0510	—
Reactive Red 120	1466	Diazo	0.0972	2.61
Reactive Violet 5	733	Monoazo	0.1770	—

Source: Smith et al. (1993).

color removals from the slack washwater as a function of ppm sorbent for TM-399 (bentonite clay modified with a quaternary ammonium surfactant), activated carbon, chitosan, and pure chitin, chitosan was found to be less effective than activated carbon. Similarly, color removal by adsorption of dyes from Navy 106 jet dye cycle effluent diluted 1 to 20 and adjusted to pH 7.0 was conducted with varying amounts of alumina, activated carbon, TM-399, and chitosan. In this test, activated carbon performed best with color changes of 4000 to 5 ADMI color units with a dosage of 2500 ppm. Chitosan was the poorest performer.

Park et al. (1995a) applied chitosan as an adsorbent for the dye Toluidine Blue O. The adsorption of Toluidine Blue O by chitosan was found to be affected by the particle size and mass of chitosan, initial dye concentration, reaction time, and pH of the solution. More dye was adsorbed with chitosan of smaller particle size and with increase in pH of the solution. When the initial ratio of dye to chitosan was more than 1:500, the adsorption of dye rapidly declined.

In a further study, Park et al. (1995b) applied chitin and chitosan prepared from red crab and squid pen as adsorbents for trapping dyes in wastewater from dyeworks. They found that chitin and chitosan were effective adsorbents for such dyes, with chitosan being more effective in dye adsorption than was chitin. In a continuous elution column experiment, the researchers claimed that 1 kg of chitosan could be used for treatment of as much as 120 L of wastewater containing 0.05% dye wasted from dyeworks at 75% efficacy, i.e., 45 g of dye absorbed/kg of chitosan.

D. Phenol- and PCB-Contaminated Wastewater

Phenol-Contaminated Wastewater. Phenols represent one of the most important classes of synthetic industrial chemicals and are often found in effluents from various manufacturing operations (Keith and Telliard 1979). Furthermore, phenols are common components of pulp and paper wastes and have been observed as groundwater contaminants. Despite the importance of treating phenol-containing wastewaters, current methods, such as physical, chemical, and microbiological treatments, are not satisfactory (Sun et al. 1992).

A two-step approach for removing phenols from wastewater has been investigated by Sun et al. (1992). In the first step, weakly adsorbable phenols are converted to quinones by the mushroom enzyme tyrosinase. The tyrosinase-generated quinones are then chemisorbed onto chitosan. This proposed approach is illustrated by the following:

Step 1: tyrosinase reaction

phenol → *o*-quinone + other intermediates

Step 2: chemisorption

o-quinone + other intermediates + sorbent → chemisorbed compounds

When mushroom tyrosinase and chitosan were added simultaneously to dilute, phenol-containing solutions, a nearly complete removal of phenols was observed.

Sun et al. (1992) mentioned three potential benefits of the two-step tyrosinase reaction–chitosan adsorption approach. As adsorption of quinones onto chitosan is very strong, the two-step approach should be more effective for removing traces of phenols from wastewater. The second benefit is that tyrosinase can react with a wide range of phenols and is less sensitive to changes in waste stream composition and strength. The final benefit is that chitosan is obtained from chitin, a waste product of the shellfish industry. However, as the mushroom enzyme is quite expensive, a less costly tyrosinase should be available for practical applications.

Payne et al. (1992) also found that a two-step tyrosinase reaction–chitosan adsorption approach could be used to selectively remove phenols from aqueous mixtures containing nonphenols (anisole and benzyl alcohol). The mushroom tyrosinase was specific for the phenol and did not react with either the anisole or the benzyl alcohol. Chitosan effectively adsorbed the tyrosinase-generated products without adsorbing nonphenols.

Removal of phenols from wastewater by soluble and immobilized tyrosinase was investigated by Wada et al. (1993). These workers found that phenols in an aqueous solution were removed after treatment with mushroom tyrosinase, with the reduction rate of phenols accelerated in the presence of chitosan. They further found that by treatment with tyrosinase immobilized on cation exchange resins, 100% of phenol was removed after 2 hr, and enzymatic activity was reduced only slightly even after 10 repeated treatments.

PCB-Contaminated Wastewater. PCBs are synthetic organic molecules widely used in various industrial sectors (e.g., plastics, electricity, lubricants, and hydraulic systems), notably to develop, combined with other organic substances, insulating fluids used in electric transformers and capacitors. PCBs always have been used in complex mixtures of congeners characterized by their different chlorine contents. PCB mixtures are sold under trade names, such as Aroclor® (Monsanto, USA), Clophen® (Bayer, Germany), and Phenoclor® (France). The Aroclor mixtures, especially Aroclor 1254 and Aroclor 1260 (respective chlorine contents, 54% and 60% by weight), have been widely used in Europe and North America (Thomé et al. 1997).

In aquatic ecosystems, PCBs are considered as priority pollutants according to their high lipophility, low water solubility, and low biodegradability. Standard methods for water purification used in water softening plants remain largely ineffective in eliminating these highly persistent toxic compounds. Up to now, expensive treatments, leading only to an incomplete elimination of PCBs such as adsorption onto activated charcoal or synthetic resins, have been used. Several investigators have observed that chitosan could perhaps fulfill the requirements of being an efficient treatment for PCB-contaminated wastewater (Thomé et al. 1992; Van Daele and Thomé 1986).

Thomé and Van Daele (1986) studied PCB adsorption capabilities of chitosan and other adsorbing agents by filtration of distilled water supplemented with 0.5 ppb PCB (Aroclor 1260) through cartridges filled with the adsorbents. As seen in Table 6, sorption abilities of chitosan were more efficient than those of other organic substances tested. Up to 84% of the PCB present in water was adsorbed by 100 mg of chitosan. C_{18} also could be used as an efficient PCB-adsorbing medium, but its high cost makes it unsuitable for application on an industrial scale.

Thomé and Van Daele (1986) also studied elimination of PCB from contaminated water (0.5 ppb) in a closed-loop system using a chitosan filter (75 g) provided with an activated charcoal filter (1.5 kg). Water circulation was achieved by a pump at a rate of 180 L/hr. It was found that complete removal of PCB from water was rapidly obtained after less than 120 hr. With an activated charcoal filter, without chitosan, concentration decreased only to 0.2 ppb after the same time.

Table 6. Efficiency of various materials to remove PCB from distilled water (100 mL) spiked with 0.5 ppb PCB.

Adsorbing material	Mean percentage of adsorption ± SD	Concentration of PCB on adsorbent ±SD (ng/g)
Chitosan (100 mg)	83.3±7.5	416±35
Chitin (100 mg)	66.1±33	330±153
Activated charcoal (200 mg)	66.6±1.2	166.4±3
Sand (200 mg)	59.2±3.6	148±9
C_{18}	97±4.1	323±14

Source: Thomé and Van Daele (1986).

Table 7. Efficiency of PCB (Aroclor 1260) binding ability of various chitosan
derivatives in a batch system (24 hr PCB/chitosan contact).

Initial PCB concentration in water (ppb)	Percent of recovery			
	CHT[a]	CHT-Glu-Red[b]	CHT-Glu[c]	CHT-BQ[d]
1	93.0	94.4	70.7	48.3
10	96.9	96.5	87.6	89.1
100	96.2	94.0	59.3	50.8
1000	39.9	82.3	73.3	58.8

[a]Unmodified chitosan.
[b]Glutaraldehyde cross-linked NaBH$_3$CN reduced chitosan.
[c]Glutaraldehyde cross-linked chitosan.
[d]Benzoquinone cross-linked chitosan.
Source: Thomé et al. (1992).

These results suggest that the highly efficient purification of PCB-contaminated
water using a chitosan filter could find favorable application in environmental situations. Van Daele and Thomé (1986) reported that filtration of the PCB-contaminated water (0.5 ppb) through chitosan was incomplete but quite sufficient to
decrease the PCB contamination to less than 1 ppm. This change was sufficient to
effectively protect fish (*Barbus barbus*) against serious metabolic diseases, such
as growth inhibition, liver and kidney volume increase, and decrease of the hemoglobin ratio in blood.

   To improve the PCB sorption property of chitosan and to elucidate the role
of the amino group in the PCB adsorption process, Thomé et al. (1992) chemically modified chitosan (CHT) by means of a cross-linking procedure with ben-
zoquinone (CHT-BQ), glutaraldehyde (CHT-Glu), and glutaraldehyde–sodium
cyanoborohydride (reductive amination, CHT-Glu-Red). In a batch system,
unmodified chitosan and the glutaraldehyde cross-linked derivative and
NaBH$_3$CN reduced chitosan (CHT-Glu-Red) were the most effective PCB adsorbents (Table 7). In a flow-through cartridge system, CHT-Glu-Red appeared to
bind PCBs most efficiently, followed by unmodified chitosan (60% of PCB
bonded on chitosan). CHT-Glu and CHT-BQ generally remained less effective.
Concerning the role of amino group in the PCB adsorption process, it was found
that the presence of the amino group (primary or secondary) in the chemically
modified chitosan was requisite for chitosan to maintain its adsorptive properties
for PCBs.

Conclusion

Researchers have focused considerable attention on development of methods for
treatment and disposal of industrial discharge waters, especially those containing
toxic wastes. In many instances, expensive treatments have led to incomplete

elimination of such toxic discharges. Thus, a need exists for more efficient and less expensive treatments of targeted wastewaters.

Chitosan is a renewable, natural, and environmentally safe biopolymer that can be obtained abundantly, and at low cost, through processing of chitinous wastes from a variety of crustacean processing industries. This biopolymer is often more effective, or more economically attractive, than other polymers such as synthetic resins, activated charcoal, cellulosic derivatives, and synthetic polyelectrolytes.

The physicochemical characteristics of chitosan can be variously affected by preparation methods and crustacean species. Therefore, the relationship between process conditions and the particular characteristics of chitosan products must be constantly monitored to achieve uniformity and proper quality control. This approach can considerably strengthen the process biotechnology supplying products of assorted grades of chitosan, selected for their particular intended use in various wastewater discharges containing real and potential hazardous compounds.

Because of its relatively low cost, combined with its high degree of efficiency, chitosan has been shown to be a competitive and powerful decontaminating agent. However, some present limitations exist for general use of chitosan in treatment of specific wastewaters. Chitosan usually is produced in a flaked or powdered form that is both nonporous and soluble in acidic media. The low internal surface area of the nonporous material limits access to interior adsorption sites, thus lowering both adsorption capacities and uptake rates of metal ions and other pollutants. The solubility of chitosan in acidic media presently limits its use as an adsorbent in low-pH wastewaters. However, chemical modifications of the biopolymer, such as cross-linking with glutaraldehyde, and the synthesis of highly porous beads of chemically cross-linked chitosan, in all likelihood will overcome usage limitations by imparting increased resistance to solubilization in acidic pH effluents, thus improving overall adsorption capacity. A broad variety of modified chitosans are available commercially.

To date, most research involved in wastewater treatments using chitosan has been conducted in batch systems. For high removal efficiency of toxic wastes, such as heavy metals, dyes, and PCBs, column operation using an immobilized chitosan is preferred to the batch system. Research should be directed toward further studies of effective column–immobilized microbe reactors for continuous removal of such toxicants. Research at Louisiana State University (Portier 1986; Portier et al. 1989) has effectively combined principles of microbial bioremediation with development of chitinous immobilization for use in effective hazardous waste detoxification and biodegradation. Successful trials with chlorinated phenol mixtures have been extended to a variety of other hazardous organic compounds. In essence, chitinous products may be utilized as an inexpensive support surface for immobilization of whole cell-adapted microbes for single and multiple toxic transformation. Regeneration of column packing materials needs to be considered to reduce wastewater treatment costs.

Additional pilot-scale studies are needed to adequately apply appropriate chitosan systems to actual waste streams from the particular "target" plant because levels of organic or inorganic compounds present in the discharge stream vary notably from plant to plant.

Summary

Research has clearly demonstrated that the biopolymer chitosan (deacetylated chitin) can be used as an effective coagulating agent for organic compounds, as a chelating polymer for binding toxic heavy metals, as well as an adsorption medium for dyes and small concentrations of phenols and PCBs present in various industrial wastewaters. In these specific applications, chitosan appears more effective than other polymers such as synthetic resins, activated charcoal, and even chitin itself. In addition, the amino group in chitosan is an effective functional group that can be altered chemically for production of other chitinous derivatives with specific useful characteristics as effective absorptive agents.

Chitosans exhibiting different physicochemical characteristics, i.e., molecular weight, crystallinity, deacetylation, particle size, and hydrophilicity, differ in their effectiveness as waste treatment agents. The specific relationship between methods and the particular crustacean species used in preparation of chitosan for wastewater treatment needs further examination. Use of bioremediation approaches, combined with immobilization of specific microorganisms on immobilized chitinous columns, is an extremely promising area of current research and actual plant operation.

Acknowledgments

We appreciate the cooperation of Dr. R. Portier of the Louisiana State University Environmental Studies Institute in supplying relevant information on his extensive bioremediation research and engineering of immobilized biopolymer–microbial reactor systems.

References

Asano T, Havakawa N, Suzuki T (1978) Chitosan applications in wastewater sludge treatment. In: Muzzarelli RAA, Pariser ER (eds) Proceedings of the First International Conference on Chitin/Chitosan. MIT Sea Grant Program, Cambridge, MA, pp 231–252.

Bough WA (1975a) Coagulation with chitosan—an aid to recovery of by-products from egg breaking wastes. Poult Sci 54:1904–1912.

Bough WA (1975b) Reduction of suspended solids in vegetable canning waste effluents by coagulation with chitosan. J Food Sci 40:297–301.

Bough WA (1976) Chitosan—a polymer from seafood wastes, for use in treatment of food processing wastes and activated sludge. Process Biochem 11(1):13–16.

Bough WA, Landes DR (1976) Recovery and nutritional evaluation of proteinaceous solids separated from whey by coagulation with chitosan. J Dairy Sci 59(11):1874–1880.

Bough WA, Landes DR (1978) Treatment of food-processing wastes with chitosan and nutritional evaluation of coagulated by-products. In: Muzzarelli RAA, Pariser ER (eds) Proceedings of the First International Conference on Chitin/Chitosan. MIT Sea Grant Program, Cambridge, MA, pp 218–230.

Bough WA, Shewfelt AL, Salter WL (1975) Use of chitosan for the reduction and recovery of solids in poultry processing waste effluents. Poult Sci 54:992–1000.

Bough WA, Wu ACM, Campbell TE, Holmes MR, Perkins BE (1978) Influence of manufacturing variables on the characteristics and effectiveness of chitosan products. II. Coagulation of activated sludge suspensions. Biotechnol Bioeng 20:1945–1955.

Coughlin RW, Deshaies MR, Davis EM (1990) Chitosan in crab shell wastes purifies electroplating wastewater. Environ Prog 9(1):35–39.

Deans JR, Dixon BG (1992) Uptake of Pb^{2+} and Cu^{2+} by novel biopolymers. Water Res 26(4):469–472.

Delben F, Gabrielli P, Muzzarelli RAA, Stefancich S (1994) Interaction of soluble chitosans with dyes in water. II. Thermodynamic data. Carbohydr Polym 24:25-30.

Eiden CA, Jewell CA, Wightman JP (1980) Interaction of lead and chromium with chitin and chitosan. J Appl Polym Sci 25:1587–1599.

Giles CH, Hassan ASA, Subramanian RVR (1958) Adsorption at organic surfaces. IV. Adsorption of sulphonated azo dyes by chitin from aqueous solution. J Soc Dyers Colour 74:682–688.

Hauer A (1978) The chelating properties of Kytex H chitosan. In: Muzzarelli RAA, Pariser ER (eds) Proceedings of the First International Conference on Chitin/Chitosan. MIT Sea Grant Program, Cambridge, MA, pp 263–276.

Holme KR, Hall LD (1991) Novel metal chelating chitosan derivative: attachment of iminodiacetate moieties via a hydrophilic spacer group. Can J Chem 69:585–589.

Hsien TY, Rorrer GL (1995) Effects of acylation and crosslinking on the material properties and cadmium ion adsorption capacity of porous chitosan beads. Sep Sci Technol 30(12):2455–2475.

Inoue K, Yamaguchi T, Iwasaki M, Ohto K, Yoshizuka K (1995) Adsorption of some platinum group metals on some complexane types of chemically modified chitosan. Sep Sci Technol 30(12):2477–2489.

Ishii H, Minegishi M, Lavitpichayawong B, Mitani T (1995) Synthesis of chitosan-amino acid conjugates and their use in heavy metal uptake. Int J Biol Macromol 17(1): 21–23.

Jha IN, Iyengar L, Prabhakara Rao AVS (1988) Removal of cadmium using chitosan. J Environ Eng 114(4):962–974.

Johnson RA, Gallanger SM (1984) Use of coagulants to treat seafood processing wastewaters. J Water Pollut Control Fed 56(8):970–976.

Jun HK, Kim JS, No HK, Meyers SP (1994) Chitosan as a coagulant for recovery of proteinaceous solids from tofu wastewater. J Agric Food Chem 42(8):1834–1838.

Kawamura Y, Mitsuhashi M, Tanibe H, Yoshida H (1993) Adsorption of metal ions on polyaminated highly porous chitosan chelating resin. Ind Eng Chem Res 32:386-391.

Keith LH, Telliard WA (1979) Priority pollutants. I: A perspective view. Environ Sci Technol 13:416–423.

Knorr D (1984) Use of chitinous polymers in food—a challenge for food research and development. Food Technol 38(1):85–97.

Koyama Y, Taniguchi A (1986) Studies on chitin. X. Homogeneous cross-linking of chitosan for enhanced cupric ion adsorption. J Appl Polym Sci 31:1951–1954.

Kurita K (1987) Binding of metal cations by chitin derivatives: improvement of adsorption ability through chemical modifications. In: Yalpani M (ed) Industrial Polysaccharides: Genetic Engineering, Structure/Property Relations and Applications. Elsevier, Amsterdam, pp 337–346.

Kurita K, Chikaoka S, Koyama Y (1988) Improvement of adsorption capacity for copper(II) ion by N-nonanoylation of chitosan. Chem Lett, pp 9–12.

Kurita K, Sannan T, Iwakura Y (1979) Studies on chitin. VI. Binding of metal cations. J Appl Polym Sci 23:511–515.

LaMer VK, Healy TW (1963) Adsorption-flocculation reactions of macromolecules at the solid-liquid interface. Rev Pure Appl Chem 13:112.

Madhavan P, Nair KGR (1978) Metal-binding property of chitosan from prawn waste. In: Muzzarelli RAA, Pariser ER (eds) Proceedings of the First International Conference on Chitin/Chitosan. MIT Sea Grant Program, Cambridge, MA, pp 444–448.

Maghami GG, Roberts GA (1988) Studies on the adsorption of anionic dyes on chitosan. Makromol Chem 189:2239–2243.

Maruca R, Suder BJ, Wightmen JP (1982) Interaction of heavy metals with chitin and chitosan. III. Chromium. J Appl Polym Sci 27:4827–4837.

Masri MS, Randall VG (1978) Chitosan and chitosan derivatives for removal of toxic metallic ions from manufacturing-plant waste streams. In: Muzzarelli RAA, Pariser ER (eds) Proceedings of the First International Conference on Chitin/Chitosan. MIT Sea Grant Program, Cambridge, MA, pp 277–287.

Masri MS, Reuter FW, Friedman M (1974) Binding of metal cations by natural substances. J Appl Polym Sci 18:675–681.

McKay G (1987) Mass transport processes for the adsorption of dyestuffs onto chitin. Chem Eng Process 21:41–51.

McKay G, Blair H, Findon A (1986) Kinetics of copper uptake on chitosan. In: Muzzarelli R, Jeuniaux C, Gooday GW (eds) Proceedings of the Third International Conference on Chitin and Chitosan, Senigallia, Italy, pp 559–565.

McKay G, Blair HS, Gardner JR (1982) Adsorption of dyes on chitin. I. Equilibrium studies. J Appl Polym Sci 27:3043–3057.

McKay G, Blair HS, Gardner J (1983) The adsorption of dyes on chitin. III. Intraparticle diffusion processes. J Appl Polym Sci 28:1767–1778.

McKay G, Blair HS, Gardner JR (1984) The adsorption of dyes onto chitin in fixed bed columns and batch adsorbers. J Appl Polym Sci 29:1499–1514.

McKay G, Blair HS, Gardner JR (1987) Two resistance mass transport model for the adsorption of acid dye onto chitin in fixed beds. J Appl Polym Sci 33:1249–1257.

McKay G, Blair HS, Gardner JG, McConvey IF (1985) Two-resistance mass transfer model for the adsorption of various dyestuffs onto chitin. J Appl Polym Sci 30:4325–4335.

McKay G, Blair HS, Hindon A (1989) Equilibrium studies for the sorption of metal ions onto chitosan. Indian J Chem 28A:356–360.

Michelsen DL, Fulk LL, Woodby RM, Boardman GD (1993) Adsorptive and chemical pretreatment of reactive dye discharge. In: Tedd DW, Pohland FG (eds) Emerging Technologies in Hazardous Waste Management III. ACS Symposium Series 518. American Chemical Society, Washington, DC, pp 119–136.

Mitani T, Fukumuro N, Yoshimoto C, Ishii H (1991) Effects of counter ions (SO_4^{2-} and Cl^-) on the adsorption of copper and nickel ions by swollen chitosan beads. Agric Biol Chem 55(9):2419.

Moore KJ, Johnson MG, Sistrunk WA (1987) Effect of polyelectrolyte treatments on waste strength of snap and dry bean wastewater. J Food Sci 52(2):491–492.

Muzzarelli RAA (1973) Natural Chelating Polymers. Pergamon Press, New York.

Muzzarelli RAA (1977) Chitin. Pergamon Press, New York.

Muzzarelli RAA, Tanfani F (1982) N-(Carboxymethyl) chitosans and N-(o-carboxybenzyl) chitosans: novel chelating polyampholytes. In: Hirano S, Tokura S (eds) Proceedings of the Second International Conference on Chitin and Chitosan, Sapporo, Japan, pp 45–53.

Muzzarelli RAA, Tubertini O (1969) Chitin and chitosan as chromatographic supports and adsorbents for collection of metal ions from organic and aqueous solutions and seawater. Talanta 16:1571–1577.

Muzzarelli RAA, Rocchetti R, Muzzarelli MG (1978) The isolation of cobalt, nickel, and copper from manganese nodules by chelation chromatography on chitosan. Sep Sci Technol 13(2):153–163.

Nair KR, Madhavan P (1982) Metal binding property of chitosan from different sources. In: Hirano S, Tokura S (eds) Proceedings of the Second International Conference on Chitin and Chitosan, Sapporo, Japan, pp 187–190.

No HK, Meyers SP (1989a) Crawfish chitosan as a coagulant in recovery of organic compounds from seafood processing streams. J Agric Food Chem 37(3):580–583

No HK, Meyers SP (1989b) Recovery of amino acids from seafood processing wastewater with a dual chitosan-based ligand-exchange system. J Food Sci 54(1):60–62,70.

Ohga K, Kurauchi Y, Yanase H (1987) Adsorption of Cu^{2+} or Hg^{2+} ion on resins prepared by crosslinking metal-complexed chitosans. Bull Chem Soc Jpn 60(1):444–446.

O'Melia CA (1972) Coagulation and flocculation. In: Weber WJ (ed) Physicochemical Processes for Water Quality Control. Wiley, New York, pp 61–109.

Park RD, Cho YY, Kim KY, Bom HS, Oh CS, Lee HC (1995a) Adsorption of Toluidine Blue O onto chitosan. Agric Chem Biotechnol 38(5):447–451.

Park RD, Cho YY, La YG, Kim CS (1995b) Application of chitosan as an adsorbent of dyes in wastewater from dyeworks. Agric Chem Biotechnol 38(5):452–454.

Payne GF, Sun WQ, Sohrabi A (1992) Tyrosinase reaction/chitosan adsorption for selectively removing phenols from aqueous mixtures. Biotechnol Bioeng 40:1011–1018.

Peniche-Covas C, Alvarez LW, Argüelles-Monal W (1992) The adsorption of mercuric ions by chitosan. J Appl Polym Sci 46:1147–1150.

Peniston QP, Johnson EL (1970) Method for treating an aqueous medium with chitosan and derivatives of chitin to remove an impurity. U.S. patent 3,533,940.

Portier RJ (1986) Chitin immobilization systems for hazardous waste detoxification and biodegradation. In: Eecles II (ed) Immobilization of Ions by Naturally Occurring Materials. Norwood, London, pp 230–243.

Portier RJ, Nelson JA, Christianson JC (1989) Biotreatment of dilute contaminated ground water using an immobilized microbe packed bed reactor. Environ Prog 8:120–125.

Randal JM, Randal VG, McDonald GM, Young RN, Masri MS (1979) Removal of trace quantities of nickel from solution. J Appl Polym Sci 23:727–732.

Rorrer GL, Hsien TY, Way JD (1993) Synthesis of porous-magnetic chitosan beads for removal of cadmium ions from waste water. Ind Eng Chem Res 32:2170–2178.

Sakaguchi T, Nakajima A (1982) Recovery of uranium by chitin phosphate and chitosan phosphate. In: Hirano S, Tokura S (eds) Proceedings of the Second International Conference on Chitin and Chitosan, Sapporo, Japan, pp 177–182.

Sandford PA, Hutchings GP (1987) Chitosan-A natural, cationic biopolymer: commercial applications. In: Yalpani M (ed) Industrial Polysaccharides: Genetic Engineering, Structure/Property Relations and Applications. Elsevier, Amsterdam, pp 363–376.

Senstad C, Almas KA (1986) Use of chitosan in the recovery of protein from shrimp processing wastewater. In: Muzzarelli R, Jeuniaux C, Gooday GW (eds) Proceedings of the Third International Conference on Chitin and Chitosan, Senigallia, Italy, pp 568–570.

Seo H, Kinemura Y (1989) Preparation and some properties of chitosan beads. In: Skjåak-Braek G, Anthonsen T, Sandford P (eds) Proceedings from the 4th International Conference on Chitin and Chitosan, Trondheim, Norway, pp 585–588.

Seo T, Hagura S, Kanbara T, Iijima T (1989) Interaction of dyes with chitosan derivatives. J Appl Polym Sci 37:3011–3027.

Shimizu Y, Kono K, Kim IS, Takagishi T (1995) Effects of added metal ions on the interaction of chitin and partially deacetylated chitin with an azo dye carrying hydroxyl groups. J Appl Polym Sci 55:255–261.

Sievers DM, Jenner MW, Hanna M (1994) Treatment of dilute manure wastewaters by chemical coagulation. Trans ASAE 37(2):597–601.

Smith B, Koonce T, Hudson S (1993) Decolorizing dye wastewater using chitosan. Am Dyest Rep 82(10):18–36.

Stefancich S, Delben F, Muzzarelli RAA (1994) Interaction of soluble chitosans with dyes in water. I. Optical evidence. Carbohydr Polym 24:17–23.

Suder BJ, Wightman JP (1983) Interaction of heavy metals with chitin and chitosan. II. Cadmium and zinc. In: Ottewill RH, Rochester CH, Smith AL (eds) Adsorption from Solution. Academic Press, London, pp 235–244.

Sun WQ, Payne GF, Moas MSGL, Chu JH, Wallace KK (1992) Tyrosinase reaction/chitosan adsorption for removing phenols from wastewater. Biotechnol Prog 8:179-186.

Thomé JP, Hugla JL, Weltrowski M (1992) Affinity of chitosan and related derivatives for PCBs. In: Brine CJ, Sandford PA, Zikakis JP (eds) Proceedings from the 5th International Conference on Chitin and Chitosan, Princeton, NJ, pp 639–647.

Thomé JP, Jeuniaux C, Weltrowski M (1997) Applications of chitosan for the elimination of organochlorine xenobiotics from wastewater. In: Goosen MFA (ed) Applications of Chitin and Chitosan. Technomic, Lancaster, PA, pp 309–331.

Thomé JP, Van Daele Y (1986) Adsorption of polychlorinated biphenyls (PCB) on chitosan and application to decontamination of polluted stream waters. In: Muzzarelli R, Jeuniaux C, Gooday GW (eds) Proceedings of the Third International Conference on Chitin and Chitosan, Senigallia, Italy, pp 551–554.

Udaybhaskar P, Iyengar L, Prabhakara Rao AVS (1990) Hexavalent chromium interaction with chitosan. J Appl Polym Sci 39:739–747.

Van Daele Y, Thomé JP (1986) Purification of PCB contaminated water by chitosan: a biological test of efficiency using the common barbel, *Barbus barbus*. Bull Environ Contam Toxicol 37:858–865.

Venkatrao B, Baradarajan A, Sastry CA (1986) Adsorption of dyestuffs on chitosan. In: Muzzarelli R, Jeuniaux C, Gooday GW (eds) Proceedings of the Third International Conference on Chitin and Chitosan, Senigallia, Italy, pp 554–559.

Wada S, Ichikawa H, Tatsumi K (1993) Removal of phenols from wastewater by soluble and immobilized tyrosinase. Biotechnol Bioeng 42:854–858.

Wu ACM, Bough WA (1978) A study of variables in the chitosan manufacturing process in relation to molecular-weight distribution, chemical characteristics and waste-treatment effectiveness. In: Muzzarelli RAA, Pariser ER (eds) Proceedings of the First

International Conference on Chitin/Chitosan. MIT Sea Grant Program, Cambridge, MA, pp 88–102.

Wu ACM, Bough WA, Holmes MR, Perkins BE (1978) Influence of manufacturing variables on the characteristics and effectiveness of chitosan products. III. Coagulation of cheese whey solids. Biotechnol Bioeng 20:1957–1966.

Yamamoto H (1984) Chiral interaction of chitosan with azo dyes. Makromol Chem 185:1613–1621.

Yang TC, Zall RR (1984) Absorption of metals by natural polymers generated from seafood processing wastes. Ind Eng Chem Prod Res Dev 23:168–172.

Manuscript received February 26, 1999; accepted April 24, 1999.

Rev Environ Contam Toxicol 163:29–112 © Springer-Verlag 2000

Blood Cholinesterases as Human Biomarkers of Organophosphorus Pesticide Exposure*

Herbert N. Nigg and James B. Knaak

Contents

*Agricultural Experiment Station Journal Series No. R-06898.
Communicated by George W. Ware.

H.N. Nigg (✉)
Citrus Research and Education Center, University of Florida, 700 Experiment Station Road, Lake Alfred, FL 33850, USA.
J.B. Knaak
Department of Pharmacology and Toxicology, School of Medicine. SUNY at Buffalo, Buffalo, NY, USA.

I. Introduction

The organophosphorus (OP) insecticides were developed before and during World War II. The history of their development has been reviewed (Holmstedt 1963; Karczmar 1970; Ursdin 1970; Koelle 1981). In 1936, Schrader synthesized paraoxon, parathion, and octamethylpyrophosphoramide (OMPA, schradan) in a search for an effective cockroach control agent (Ursdin 1970). Parathion use in agriculture began after World War II. In 1949, a mixer loader was killed by parathion in Lake Placid, FL (Griffiths et al. 1951). Monitoring red blood cell acetylcholinesterase (RBC AChE) of exposed workers was begun in 1950 in the Florida citrus industry (Griffiths et al. 1951), perhaps the first use of human blood esterase monitoring in agriculture.

With the introduction of organophosphorus pesticides (OPs) into American agriculture, there has been considerable interest in their chemistry, mode of action, and toxicity. Numerous articles have been written and published in scientific journals concerning the manner in which they inhibit esterases, particularly acetylcholinesterase (AChE), located in nerve tissue and responsible for removing acetylcholine (ACh) at neuromuscular junctions (synapses). The inhibitory action of OPs on AChE appears to be largely responsible for their toxicity toward invertebrates and vertebrates. In many cases, the action of the OPs may be purely anticholinergic, although numerous studies have been carried out to determine whether these compounds are neurotoxic (producing behavioral and pathological changes), are carcinogenic, or adversely affect development or reproduction. Several OPs have been shown to produce a delayed-onset neuropathy when administered acutely and at a high dose to adult hens. This neuropathy is believed to be caused by the inhibition of an esterase in nerve tissue known as neurotoxic esterase (NTE). The OPs do not seem to produce adverse health effects in humans at tissue levels producing little or no RBC AChE inhibition.

In recent years, U.S. Environmental Protection Agency (USEPA) scientists have reexamined the relationship between OPs and AChE inhibition as part of their review of the use of RBC AChE in worker monitoring programs and in determining reference doses (RfDs) and reference concentrations (RfCs). USEPA scientists determined that the relationship between dose and RBC AChE inhibition was poor and could not be used for determining RfDs or RfCs. The USEPA also questioned the usefulness of RBC AChE in worker monitoring programs. Monitoring of blood esterases for OP pesticide exposure was proposed as part of the USEPA's Farm Worker Protection Standards. Blood esterase monitoring was not part of the final rule and is not recommended by the USEPA as a means to define risk. The USEPA "recognizes the possibility that new analyses may provide additional information on the biological significance of cholinesterase inhibition in red blood cells" (Pesticide & Toxic Chemical News 1991). The USEPA now takes a regulatory approach that places the risk from OPs together in a cumulative risk assessment. This action may become a common approach to regulating pesticides that display a common mechanism of toxicity (Mileson et al. 1998).

Worker blood esterase monitoring technology is focused on the measurement of RBC AChE and plasma cholinesterase (butyrylcholinesterase, BChE) activity. One or both of these classes of blood enzymes may be used to monitor workers exposed to OPs (Gage 1967; Hayes 1983; Peakall 1992). This review was written to address the issues raised by health and regulatory scientists on the use of OP–AChE inhibition data in the development of RfCs and RfDs and standards (i.e., worker reentry times, etc.) for agricultural worker health and safety programs. A review of this nature cannot be made without a discussion of the structure and metabolism of OPs in relationship to their antiesterase activity. Casida (1956), O'Brien (1960), and Fukuto (1971) have published extensive reviews on OP structure, metabolism and antiesterase activities. Knaak et al. (1989) reviewed the field reentry problem of agricultural workers. Wilson et al. (1997, 1998) reviewed cholinesterase (ChE) inhibition and the problems associated with measuring AChE activity in blood drawn from animals and humans exposed to OPs. We review the work of these authors and others as it pertains to the overall problem of using blood AChE and BChE inhibition as a measure of toxic response to OPs.

We present what we believe is the best blood esterase monitoring program for practical use based on the information available.

II. Nature of Organophosphorus Pesticides

A. Structure

The physicochemical properties of the OP pesticides depend upon the substituent groups about the pentavalent phosphorus atom. Seven types of OP insecticides were produced and registered for use in the United States. They are represented by phosphates (O,O-P=O(O)), phosphorothionates (O,O-P=S(O)), phosphorothiolates (O,O-P=O(S)), phosphorodithioates (O,O-P=S(S)), phosphoramidates (O,O-P=O(NHR)), phosphoramidothioates (O,O-P=S(NHR)), and phosphonates (O,R-P=O(O)). The phosphorothionates and phosphorothiolates are commonly grouped together and designated as phosphorothioates. The phosphorothioates (e.g., parathion) and phosphorodithioates (e.g., azinphosmethyl, dialifor, methidathion, dimethoate, and phosalone) are metabolized in the body and on foliage to produce more toxic products called oxons (P=O(S to O)). The oxons are better inhibitors of AChE, and their formation on plant foliage is the principal reason for the field reentry poisonings in California (Knaak et al. 1989). The analytical catalog of Riedel-de Haen Aktiengesellschaft lists 73 OPs and OP metabolites. Many of these OPs were registered by USEPA and used at one time or another for noncrop and crop protection. Approximately 41 OPs are currently registered for use in the U.S. Table 1 gives the common name, chemical name, and structure of currently registered pesticides and other OP compounds mentioned in this review. The basic structure of these pesticides is shown in Fig. 1, p. 37.

Table 1. Chemical identification of USEPA registered pesticides and other organophosphorus compounds mentioned in text.

Pesticide	Chemical structure
Dialkylvinylphosphates	
[a]**Tetrachlorfenvinphos**, *O,O*-diethyl-(*E*)-2-chloro-1-(2,4,5-trichlorophenyl) vinyl phosphate CAS 22248-79-9, MW = 365.96	
[a]**Dichlorvos** (DDVP), *O,O*-dimethyl-(2,2-dichlorovinyl) phosphate CAS 62-73-7, MW = 220.98	
[a]**Mevinphos**, *O,O*-dimethyl-(2-methoxy-carbonyl-1-methylvinyl) phosphate CAS 7786-34-7, MW = 224.15	
[a]**Monocrotophos**, *O,O*-dimethyl-cis-1-methyl-2-(methylcarbamoyl) vinyl phosphate CAS 6923-22-4, MW = 223.17	
[a]**Dicrotophos**, 3-(dimethoxyphosphinyloxy) *N,N*-dimethylisocrotonamide CAS 141-66-2, MW = 237.19	
Phosphamidon, *O,O*-dimethyl-2-chloro-2-diethylcarbamoyl-1-methyl vinyl phosphate CAS 13171-21-6, MW = 299.69	
Dialkyl aryl phosphates	
Paraoxon-methyl, *O,O*-dimethyl-*O*-(4-nitrophenyl) phosphate CAS 950-35-6 MW = 247.14	
Paraoxon, *O,O*-diethyl *O*-(4-nitrophenyl) phosphate CAS 311-45-5, MW = 275.2	
Dialkyl alkyl phosphates	
[a]**Dibrom** (Naled), *O,O*-dimethyl *O*-(1,2-dibromo-2,2-dichloroethyl) phosphate CAS 300-76-5, MW = 380.78	

Table 1. (Continued).

Pesticide	Chemical structure

Dialkyl aryl phosphorodithioates

[a]**Azinphosmethyl**, *O,O*-dimethyl-
S-(3,4-dihydro-4-oxobenzo[*d*]-[1,2,3]-triazin-3-
yl) methyl) phosphorodithioate
CAS 86-50-0, MW = 317.32

Dialifor, *O,O*-diethyl-
S-(2-chloro-1-phthalimidoethyl)
phosphorodithioate
CAS 10311-84-9, MW = 393.84

Dioxathion, *O,O,O′,O′*-tetraethyl-
S,S′-(1,4-dioxane 2,3-diyl)
di-(phosphorodithioate)
CAS 78-34-2, MW = 456.52

[a]**Fonophos**, *O*-ethyl *S*-phenyl
ethylphosphorodithioate
CAS 944-22-9, MW = 246.32

[a]**Methidathion**, *O,O*-dimethyl-
S-[2,3-dihydro-5-methoxy-2-oxo-1,3,4-
thiadiazol-3-methyl] phosphorodithioate
CAS 950-37-8, MW = 302.32

Phosalone, *O,O*-diethyl
S-(6-chloro-benzoxazolon-3-yl)-methyl
phosphorodithioate
CAS 2310-17-0, MW = 367.8

[a]**Phosmet**, *O,O*-dimethyl-S-phthalimidomethyl
phosphorodithioate
CAS 732-11-6, MW = 317.31

[a]**Bensulide**, *O,O*-diisopropyl-
S-(2-phenylsulfonylaminoethyl)
phosphorodithioate
CAS 741-58-2, MW = 397.50

Phosphonate

[a]**Trichlorfon**, *O,O*-dimethyl-
(2,2,2-trichloro-1-hydroxyethyl)-phosphonate
CAS 52-68-6, MW = 257.44

34 H.N. Nigg and J.B. Knaak

Table 1. (Continued).

Pesticide	Chemical structure

Dialkyl carboxylic acid amide/ester phosphorodithioate

[a]**Dimethoate**, *O,O*-dimethyl
S-(*N*-methylcarbamoylmethyl)
phosphorodithioate
CAS 60-51-5, MW = 229.25

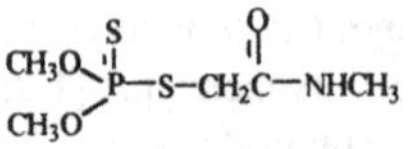

[a]**Malathion**, *O,O*-dimethyl S-[1,2-di-
(ethoxycarbonyl) ethyl] phosphorodithioate
CAS 121-75-5, MW = 330.35

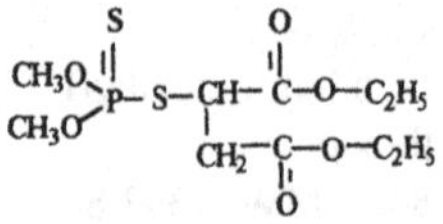

Dialkylphosphoramidothioates/amidates

[a]**Acephate**, *O, S*-dimethyl-
N-acetylphosphoramidothioate
CAS 30560-19-1, MW = 183.16

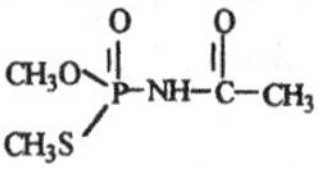

[a]**Isofenphos**, *O*-ethyl-
O-2-isopropoxy-carbonylphenyl isopropyl
phosphoramidothioate
CAS 25311-71-1, MW = 345.39

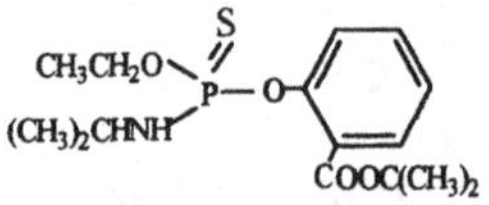

[a]**Fenamiphos**, *O*-ethyl 4-(methylthio-)*m*-tolyl
isopropyl phosphoramidate
CAS 22224-96-6, MW = 303.30

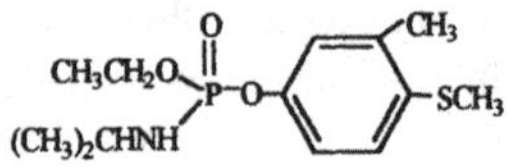

[a]**Propetamphos**, *O*-methyl,
(*E*)-*O*-2-isoproxycarbonyl-1-methylvinyl ethyl
phosphoramidothioate
CAS 31218-83-4, MW = 281.31

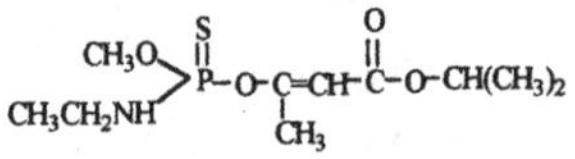

[a]**Methamidophos**, *O, S*-dimethyl
phosphoramidothioate
CAS 10265-92-6, MW = 141.12

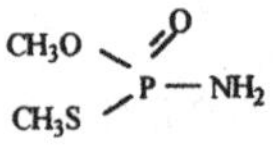

Phosphorothionates

[a]**Chlorpyrifos-methyl**, *O,O,*-dimethyl-
O-(3,5,6-trichloro 2-pyridyl) phosphorothioate
CAS 5598-13-0. MW = 322.53

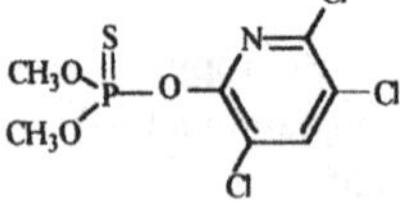

[a]**Chlorpyrifos**, *O,O*-diethyl-
O-(3,5,6-trichloro-2-pyridyl) phosphorothioate
CAS 2921-88-2, MW = 350.58

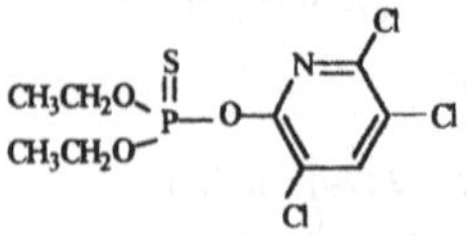

Chlorothiophos, *O,O*-diethyl-
O-[2,5-dichloro-4-(methylthio) phenyl]
phosphorothioate
CAS 60238-56-4, MW = 361.25

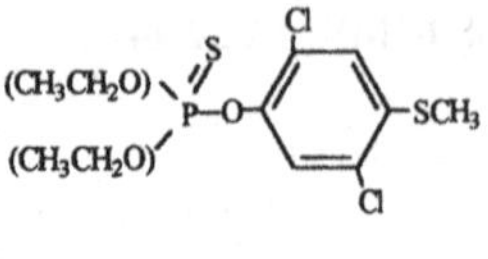

Table 1. (Continued).

Pesticide	Chemical structure
[a]**Oxydemetonmethyl** (Metasystox R), *O,O* dimethyl *S*-[2-(ethylsulfinyl)ethyl] phosphorothioate CAS 8022-00-2, MW = 213.91	
Demeton, *O,O*-diethyl *O*-[2-(ethylthio)ethyl] phosphorothioate—mixture with *O,O*-diethyl *S*-[2-(ethylthio)ethyl] phosphorothioate CAS 80 65-48-3, MW = 258.34	
[a]**Diazinon**, *O,O*-diethyl- *O*-(2-isopropyl-6-methyl-4-pyrimidinyl) phosphorothioate CAS 333-41-5, MW = 304.34	
[a]**Pirimiphosmethyl**, *O,O*,-dimethyl- *O*-2-diethylamino-6-methyl pyrimidine-4-yl phosphorothioate CAS 29232-93-7, MW = 305.33	
[a]**Tebupirimphos** (Phostebupirim), *O*-ethyl, *O*-(1-methylethyl)- *O*-[2-(1,1-dimethylethyl)-5-pyrimidinyl] phosphorothioate CAS 96182-53-5, MW = 317.80	
[a]**Famophos**, *O,O*-dimethyl- *O*-*p*-(dimethylsulfamoyl) phenyl phosphorothioate CAS 52-85-7, MW = 325.33	
[a]**Coumaphos**, *O,O*-diethyl- *O*-(3-chloro-4-methyl-7-coumarinyl) phosphorothioate CAS 56-72-4, MW = 362.76	
[a]**Fenthion**, *O,O*-dimethyl-*O*-(3-methyl-4-methyl thiophenyl phosphorothioate CAS 55-38-9, MW = 278.32	
[a]**Parathion-methyl**, *O,O*-dimethyl *O*-4-(nitrophenyl)-phosphorothioate, CAS 298-00-0, MW = 263.20	
[a]**Parathion**, *O,O*-diethyl *O*-(4-nitrophenyl) phosphorothioate CAS 56-38-2, MW=291.26	

Table 1. (Continued).

Pesticide	Chemical structure
[a]**Fenitrothion**, *O,O*-dimethyl-*O*-(3-methyl-4-nitrophenyl) phosphorothioate CAS 122-14-5, MW = 277.23	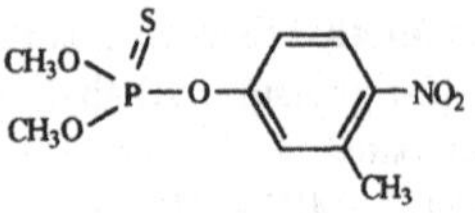

Dialkylthioalkylphosphorodithioates/trithioates

[a]**Terbufos**, *O,O*-diethyl-*S*-[tert-butyl-thiomethyl] phosphorodithioate CAS 13071-79-9, MW = 288.42	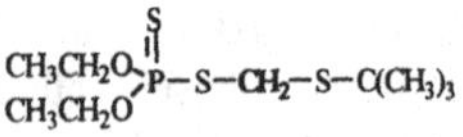
[a]**Disulfoton**, *O,O*-diethyl-*S*-[2-(ethylthio)-ethyl] phosphorodithioate CAS 298-04-0, MW = 274.41	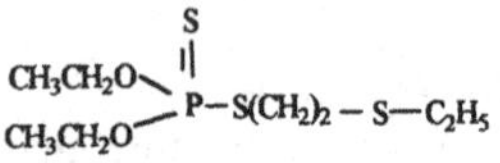
[a]**Phorate**, *O,O*-diethyl *S*-[(ethylthio)methyl] phosphorodithioate CAS 298-02-2, MW = 260.36	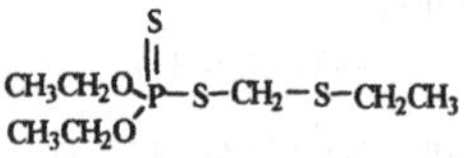
[a]**Ethoprop**, *O*-Ethyl-*S,S*-dipropyl phosphorodithioate CAS 13194-48-4, MW = 242.3	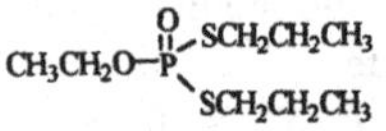
[a]**DEF**, *S,S,S*-tributylphosphoro-trithioate CAS 78-48-8, MW = 314.50	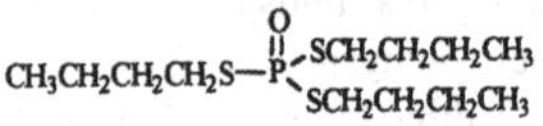

Diphosphate esters

[a]**Ethion**, *O,O,O',O'*-tetraethyl *S,S'*-methylenediphosphorodithioate CAS 563-12-2, MW = 384.46	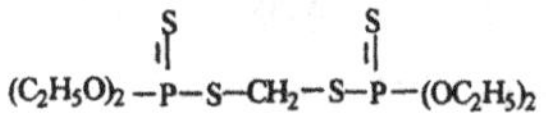
[a]**Temephos**, *O,O,O',O'*-tetramethyl-*O,O'*-(thiodi-*p*-phenylene) diphosphorothioate CAS 3383-96-8, MW = 466.46	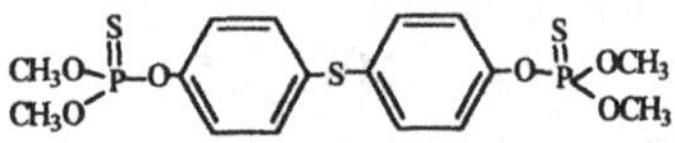
TEPP, *O,O,O',O'*-tetraethylpyrophosphate CAS 107-49-3, MW = 290.19	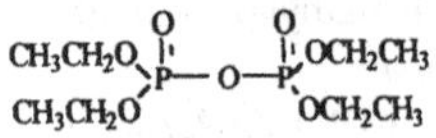
[a]**Sulfotep**, bis-*O,O*-diethylphosphorothionic anhydride CAS 3689-24-5, MW = 322.31	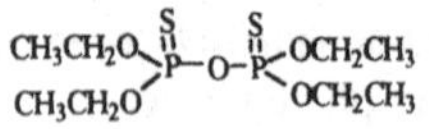
OMPA, N,N'-diisopropyl phosphorodiamidic anhydride MW = 333.70	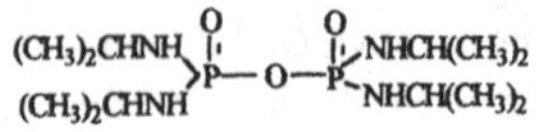

Phosphorofluoridates

DFP, (bis(1-methylethyl) phosphorofluoridate) CAS 55-91-4, MW = 184.15	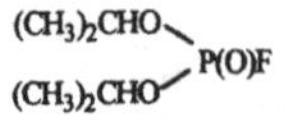

Table 1. (Continued).

Pesticide	Chemical structure
[b]Soman, (1,2,2-trimethylpropyl methylphosphonofluoridate) CAS 96-64-0, MW = 182.17	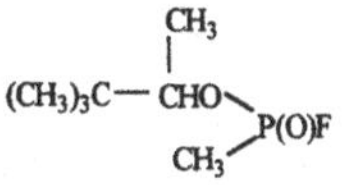 $(CH_3)_3C-CHO$ CH_3 $P(O)F$ CH_3
[b]Sarin, (1-methylethyl methylphosphonofluoridate) CAS 107-44-8, MW = 140.09	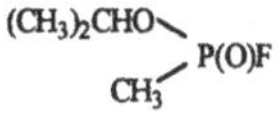 $(CH_3)_2CHO$ CH_3 $P(O)F$

[a]Active pesticide ingredient registered for use by Environmental Protection Agency.
[b]Chemical used as nerve gas in chemical warfare.
Additional active pesticides registered for use by USEPA:
chlorethoxyfos, *O,O*-diethyl *O*-(1,2,2-tetrachloroethyl) phosphorothioate, CAS 54593-83-8, MW = 336.0.
isazofos, *O,O*-dimethyl *O*-[5-chloro-1-(1-methylethyl)-1*H*-1,2,4-triazol-3-yl] phosphorothioate, CAS 42509-80-8.
profenofos, *O*-(4-bromo-2-chlorophenyl) *O*-ethyl *S*-propyl phosphorothioate, CAS 41198-08-7, MW = 373.63.

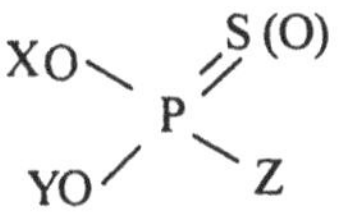

$$XO\diagdown_{}\overset{S\,(O)}{\overset{\diagup}{\underset{P}{}}}$$
$$YO\diagup\diagdown Z$$

Fig. 1. Structure of organophosphorus insecticide. X and Y are alkyl groups, and Z represents an alkyl or aryl leaving group.

B. Stability and Activity

The OPs are neutral esters, biologically active against insects, and readily hydrolyze in acid or basic medium to give their respective alkyl phosphates and leaving groups (aryl or alkyl). The relationship between their rate of hydrolysis and structure has been extensively studied. Compounds possessing dimethyl $(CH_3O)_2$ and diethyl $(CH_3CH_2O)_2$ groups are more readily hydrolyzed than those with isopropyl or larger groups. Substitution of P=O for P=S enhances the rate of hydrolysis, as does isomerization of P(S)-O-R thionate for P(O)S-R thiolate (O'Brien 1960) in compounds such as demeton. The oxidation of mercapto sulfur in the case of the thiono form of demeton to sulfoxide (S(O)R) or a sulfone (-S(O)(O)R) also leads to a decrease in stability. In the case of the thioloisomer the effect is positive but somewhat smaller. Generally speaking, hydrolyzability is directly related to anticholinesterase activity, but not always as is the case with dimethyl, diethyl, and diisopropyl phosphorofluoridates in which hydrolyzability decreases and anticholinesterase properties increase. The relationship between structure, stability, and activity is complex involving not only OP structure, but the active site of the enzyme(s) as well.

A number of quantitative structure–activity relationship (QSAR) models (Schüürmann, 1992) for predicting *in vitro* or *in vivo* OP inhibition of ChEs have been developed since the early work of Fukuto and Metcalf (1956). Raabe et al. (1994) recently proposed the development of a QSAR model for predicting the dermal toxicity of OPs using dermal dose–ChE response data (ED_{50} values). The cross-validation tests used by Enslein et al. (1998) to develop the TOPKAT® LD_{50} model were to be used for this model as well. The use of skin permeability data (K_p values) on OPs in the dermal toxicity model was suggested by Raabe et al. (1994) as a means for improving the predictability of the model.

III. Metabolism of Organophosphorus Pesticides

In biological systems, OPs undergo metabolism by four general reaction classes: (1) reactions involving the P-450 and mixed-function oxidases (MFO), (2) reactions involving hydrolases ('A'- and 'B'-esterases), (3) transferase reactions (glucuronic and sulfuric acid and glutathione) and (4) miscellaneous reactions (Dauterman 1971). The scientific literature contains metabolic data on a large number of compounds; a few examples are provided in Table 2 to illustrate these reactions and the enzymes involved.

A. Metabolic Enzymes and Their Reactions

1. *Oxidative Metabolism.* Liver cytochrome P-450 isozymes fortified with the reduced coenzyme, nicotinamide adenine dinucleotide phosphate (NADPH), are capable of carrying out oxidative desulfuration (P=S to P=O), N-dealkylation, O-dealkylation, O-dearylation, thioether and side chain oxidation, as indicated in Table 2 (Hodgson and Levi 1992, 1994). The phosphorothioates and phosphorodithionates readily undergo desulfuration (P=S to P=O) to form toxic oxons. Parathion, methyl parathion, malathion, dimethoate, diazinon, fenitrothion, methidathion, and isofenphos are examples of compounds that undergo this reaction. OPs possessing dialkyl carboxylic acid amide/ester groups undergo oxidative N-dealkylation (i.e., dicrotophos; Table 2, schradan, dimethoate, famphur, and phosphamidon) to less toxic compounds. Dicrotophos undergoes this reaction to form monocrotophos (Azodrin®), a commercial OP. P-450 isozymes are also capable of removing one alkyl group from a phosphate such as chlorfenvinphos, resulting in a dramatic loss in toxicity. This oxidative O-dealkylation by P-450 isozymes to form des methyl or des ethyl compounds is shown in Table 2 for chlorfenvinphos and dimethylnaphthyl phosphate. The cleavage of the P-O-aryl bond by an oxidative mechanism has been reported for parathion, methyl parathion, fenitrothion, and diazinon. Sultatos and Murphy (1983) demonstrated the conversion of chlorpyrifos to chlorpyrifos oxon and hydrolysis of the oxon to yield 3,5,6-trichloro-2-pyridinol. Studies with isofenphos resulted in desulfuration without hydrolysis (Knaak et al. 1993b). Cleavage of the P-S-aryl bond was reported for fonophos (a phosphonothiolothioate) and for malathion and azinphosmethyl (Hodgson and Levi 1994) (this reaction is not shown in Table 2). Thioether-containing OPs such as disulfoton, demeton, fensulfothion, or phorate can be activated to more potent anticholinesterase agents by P-450 isozymes

Table 2. Metabolism and enzymes.

P-450 Oxidative desulfuration

Parathion:

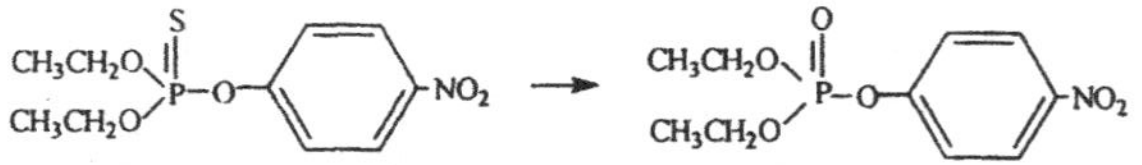

methyl parathion → methyl paraoxon   Gage (1953) / Metcalf and March (1953)

methyl parathion → methyl paraoxon   Hollingworth et al. (1967)

Malathion → malaoxon   O'Brien (1957)

dimethoate → dimethoate oxon   Dauterman et al. (1960)

fenitrothion → fenitrothion oxon   Hollingworth et al. (1967)

methidathion → methidathion oxon   Bull (1970)

isofenphos → isofenphos oxon   Knaak et al. (1996)

P-450 Oxidative N-dealkylation

dicrotophos:

schradan → des N-methyl schradan   Menzer and Casida (1965) / Bull and Lindquist (1964, 1966) / Linquist and Bull (1967)

dimethoate → des N-methyl dimethoate   Spencer et al. (1957)

     Sanderson and Edson (1964) / Lucier and Menzer (1968, 1970)

Table 2. (Continued).

famphur → des N-methyl famphur	O'Brien et al. (1965)
phosphamidon → des N-ethylphosphamidon	Bull et al. (1967)
	Clemmons and Menzer (1968)

P-450 Oxidative O-dealkylation

chlorfenvinphos: Donninger et al. (1967)

naphthyl phosphate: Hutson et al. (1969)

Table 2. (Continued).

P-450 Oxidative O-dearylation

parathion:

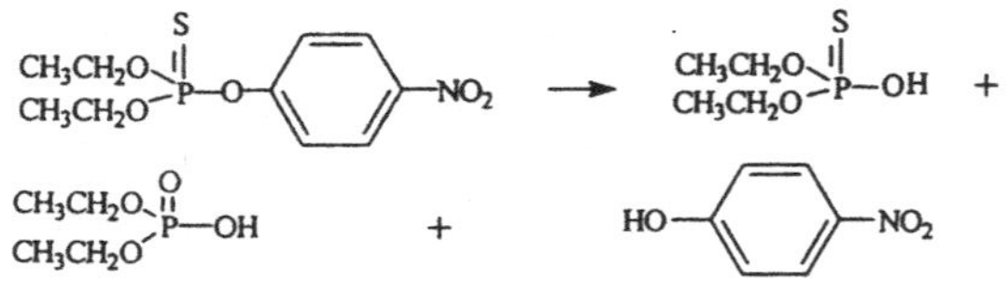

	References
	Fukunaga (1967)
	Neal (1967a,b)
	Nakattsugawa and Dahm (1967)
	El Bashir and Openoorth (1969)
	Yang et al. (1971a,b)

P-450 Thioether oxidation

disulfoton:

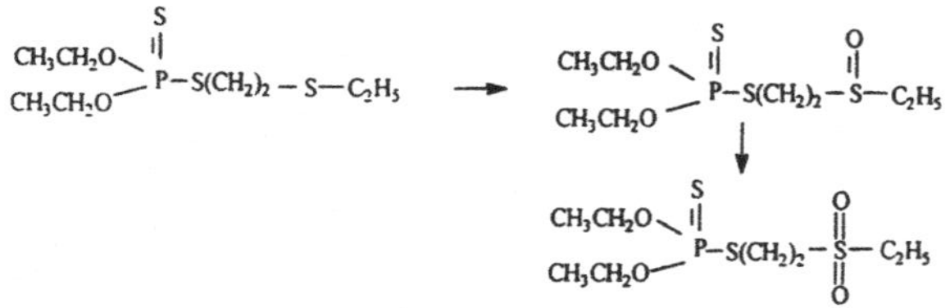

Metcalf et al. (1957)
Bull (1970)

Hodgson et al. (1995)

demeton → demeton sulfoxide → demeton sulfone

Fukuto et al. (1955)
Fukuto et al. (1956)

phorate → phorate sulfoxide → phorate sulfone

Bowman and Casida (1957)
Metcalf et al. (1957)

fensulfothion → fensulfothion sulfoxide → fensulfothion sulfone

Benjamini et al. (1959a,b)

Table 2. (Continued).

P-450 Side group oxidation

fenitrothion:

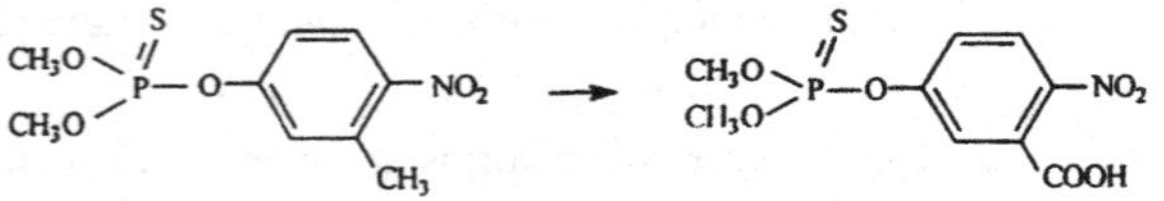

Douch et al. (1968)

diazinon:

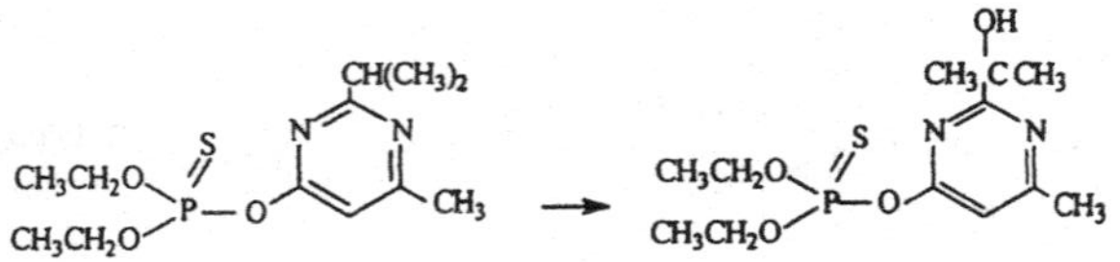

Pardue et al. (1970)
Mücke et al. (1970)

Phosphotriesterases/"A" esterases

DFP

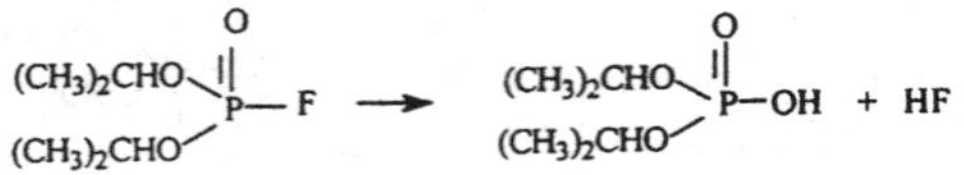

Mounter (1956)
Cohen and Warringa (1957)

diazoxon:

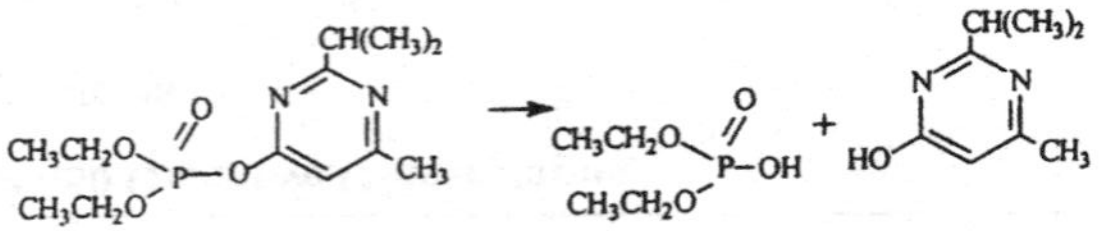

Matsumura and Hogendijk (1964)

Table 2. (Continued).

paraoxon

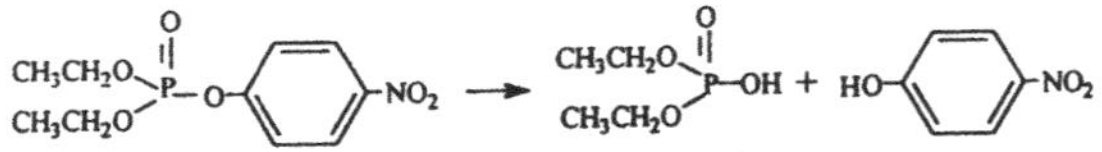

CH_3CH_2O—P(=O)—O—⟨C$_6$H$_4$⟩—NO_2 → CH_3CH_2O—P(=O)—OH + HO—⟨C$_6$H$_4$⟩—NO_2
Aldridge (1953a)
Main (1960a,b)

Carboxylesterases/"B" esterases

malathion:

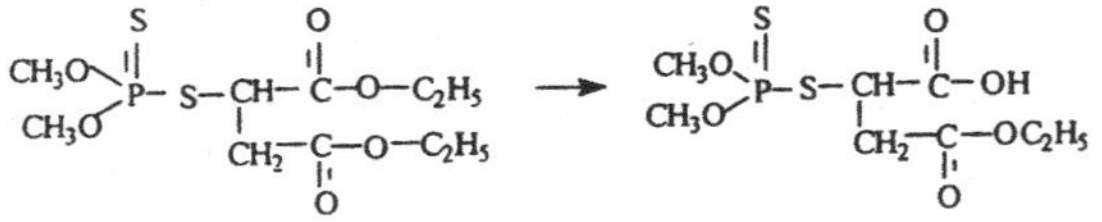

CH_3O—P(=S)—S—CH—C(=O)—O—C_2H_5 → CH_3O—P(=S)—S—CH—C—OH
$\quad\quad\quad\quad\quad\quad CH_2$—$C$(=O)—$O$—$C_2H_5$ $\quad\quad\quad CH_2$—C(=O)—O—C_2H_5

Cook and Yip (1958)
Cook et al (1958)

Amidase/"B" esterases

dimethoate:

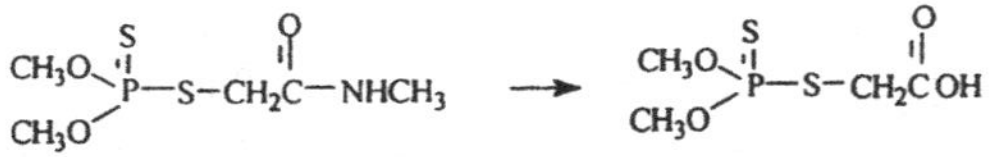

CH_3O—P(=S)—S—CH_2C(=O)—$NHCH_3$ → CH_3O—P(=S)—S—CH_2COH(=O)
Menzie (1969)
Uchida et al. (1964)
Uchida and O'Brien (1967)

Transferases

Glucuronic acid conjugation: (phenol)

⟨C$_6$H$_5$⟩—OH + UDPG → ⟨C$_6$H$_5$⟩—O—Glucuronic acid
Smith and Williams (1966)
Whetstone et al. (1966)

Table 2. (Continued).

Sulfuric acid conjugation: (phenol)

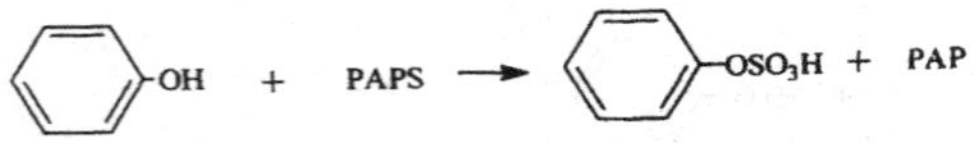

$$\text{OH} + \text{PAPS} \rightarrow \text{OSO}_3\text{H} + \text{PAP}$$

Parke (1968)

Glutathione conjugation

S-alkyl

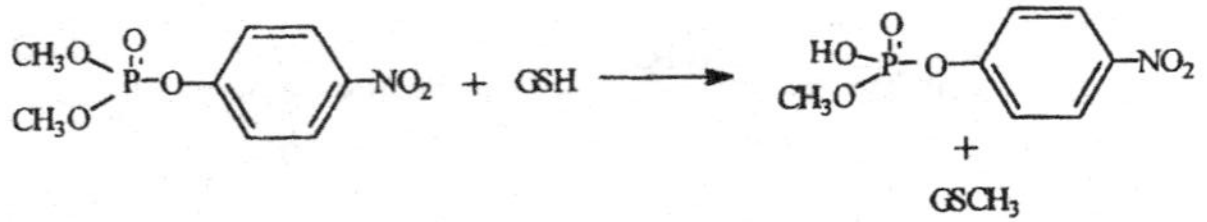

Hollingworth (1969, 1970)

S-aryl

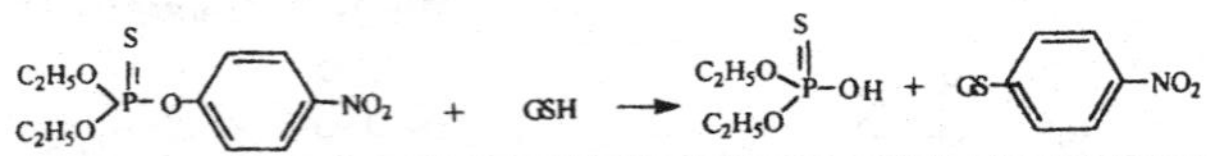

Hollingworth et al. (1973)

with the formation of sulfoxide or sulfone derivatives. Fenitrothion and diazinon undergo P-450-catalyzed side group oxidations shown in Table 2.

Cytochrome P-450 isozymes also catalyze ring hydroxylation, deamination, alkyl and N-hydroxylation, and N-oxide formation (Eto 1974; Matsumura 1975). Knaak et al. (1993b) developed V_{max}, K_m values for the desulfuration and N-deisopropylation of isofenphos by P-450 enzymes. The values were used in the development of a physiologically based pharmacokinetic/pharmacodynamic model (Knaak et al. 1996). Wallace and Dargan (1987) developed V_{max} and K_m values for the desulfurization and hydrolysis of parathion. V_{max} and K_m values for desulfurization were similar to those published by Knaak et al. (1993b) for isofenphos. A kinetic analysis of the desulfurization (activation) and dearylation (detoxification) of parathion and chlorpyrifos (V_{max}, K_m values) was carried out by Ma and Chambers (1994) in the rat. Rat liver microsomes (P-450 enzymes) had a higher capacity to activate parathion and a lower capacity to detoxify parathion. Differences in these activities appear to be related to the acute toxicities of parathion compared to chlorpyrifos.

2. *Hydrolysis by esterases.* A series of hydrolases (phosphorylphosphatases, arylesterases, carboxylesterases, carboxylamidases, and phosphotriesterses) are capable of hydrolyzing OPs (Table 2). In 1953, Aldridge proposed a classification involving the interaction of esterases with OPs (Aldridge 1953a,b). According to Aldridge, 'A'-esterases hydrolyze OPs (i.e., oxons), while 'B' esterases are inhibited by OPs. OPs (oxons) act as substrates for both 'A'- and 'B'-esterases with 'B'-esterases binding a phosphoryl group to a serine residue at the active site of the enzyme (Fig. 2). [See Section V on inhibition kinetics.]

AChE (EC 3.1.1.7), ChE (EC 3.1.1.8), carboxylesterase, and carboxylamidases are 'B' esterases, while paraoxonase and diisopropyl fluorophosphate (DFP) DFPase are 'A' esterases. Paraoxonases are found in blood serum and in hepatic microsomes and are Ca^{2+} dependent, while DFPase is found in serum, hydrolyzes the P–F bond, and is dependent upon Mn^{2+} and Ca^{2+}. Carboxylesterases are present in rat liver microsomes and are active against carboxylesters (EC 3.1.1.1) (Reiner et al. 1989). The mechanisms involved in the hydrolysis of OP compounds by 'A'-esterases are still unknown. The catalytic center amino acid is also unknown. Kinetic evidence exists for the presence of a Michaelis complex between the enzyme and OP, but no evidence exists for any other intermediate and the order of release of products is unknown. 'A'-esterases require OP compounds to be fully esterified to be a substrate. The substituents about the phosphorus can vary widely, with hydrolysis taking place on the P–O, P–F, or P–CN bonds. Some evidence exists for hydrolysis of P–S bonds. According to Chambers et al. (1994), the 'A'-esterase-mediated hydrolysis of chlorpyrifos-oxon but not of paraoxon occurred at biologically relevant nanomolar (nM) concentrations. Zimmerman et al. (1989) reported a K_m value of 0.64 mM for rabbit serum paraoxonase. Specific activities of 0.056 mM min^{-1} mg^{-1} of protein were reported for paraoxon. For soman, K_m values of 0.5 mM and V_{max} values of 14.0 nM min^{-1} mL^{-1} of a 1.5% liver homogenate have been reported for phosphorylphosphatases (de Jong et al. 1989).

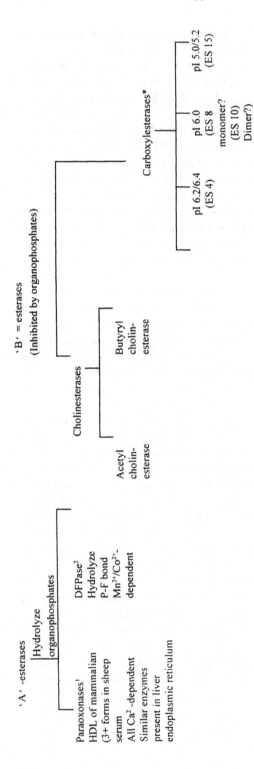

Fig. 2. Esterase classification (Walker 1989).

3. *Conjugation of Leaving Group*. Glucuronic acid is transferred to phenolic, hydroxylamino, and alcoholic hydroxyl groups, carboxyl groups, amino and imino groups, and sulfhydryl groups (Smith and Williams 1966). The transfer is catalyzed by microsomal glucuronyl transferase and requires the cofactor uridine diphosphate glucuronic acid (UDPGA). Sulfate conjugation occurs primarily with aromatic amino groups, phenolic hydroxyl groups, and aliphatic alcoholic hydroxyl groups (Parke 1968). The sulfate-transferring enzyme aryl sulfotransferase is involved along with a cofactor, adenosine-3'-phosphate-5'-phosphosulfate. Conjugation produces water soluble compounds that are readily excreted in urine. In Table 2 phenol is shown being conjugated to glucuronic and sulfuric acids.

4. *Glutathione Transferases*. Glutathione transferases (postmicrosomal fraction of liver) catalyze the transfer of O-alkyl and O-aryl groups to glutathione. Reactions involving dimethyl-substituted phosphate and phosphorothioate esters such as dichlorvos, methyl paraoxon, methyl parathion, fenitrothion, and azinphosmethyl have been reported (Hollingworth 1970, 1971; Fukami and Shishido 1966; Fukunaga et al. 1969). O-alkyl transfers yield *S*-methyl glutathione and monodesmethyl products. Hollingworth et al. (1973) demonstrated the formation of *S*-*p*-nitrophenylglutathione from parathion.

B. Metabolic Pathways

The individual metabolic reactions and enzymes involved in the biotransformation of OPs in the animal body were discussed in the previous section. The following metabolic pathways for isofenphos, parathion, and malathion describe all the known metabolic intermediates and excretable end products formed on these pesticides in the intact animal. Isofenphos and malathion are of interest because these compounds have carboxylester groupings capable of being hydrolyzed by the action of carboxylesterases. Parathion, malathion, and isofenphos all possess a P=S bond capable of being converted to P=O by the action of P-450 isozymes. All three pesticides are metabolized to alkyl phosphates that are readily eliminated in urine.

1. *Isofenphos*. Figure 3 gives the combined metabolic pathway for isofenphos in the rat, guinea pig, and dog (Knaak et al. 1996). Isofenphos is first metabolized by P-450 enzymes to isofenphos oxon, des *N*-isopropyl isofenphos, and des *N*-isopropyl isofenphos oxon (reactions 1, 2, 3, and 4). The two oxons are inhibitors of AChE, BChE, and carboxylesterase. Isofenphos and des *N*-isopropyl isofenphos are metabolized by carboxylesterase to give their respective carboxylic acids (reactions 5 and 6), while the two oxons are primarily metabolized by OP hydrolases to give their respective alkyl phosphates (AP1 and AP2) and isopropyl salicylate (IPS) (reactions 12 and 13). Unlike parathion, isofenphos is not readily hydrolyzed by P-450 enzymes. des *N*-isopropyl isofenphos is hydrolyzed by OP hydrolases to its respective alkyl phosphate (AP4) (reaction 11). Phosphoamidases remove amino groups to form des amino isofenphos and des amino isofenphos oxon (reactions 9 and 10). These compounds may be further hydro-

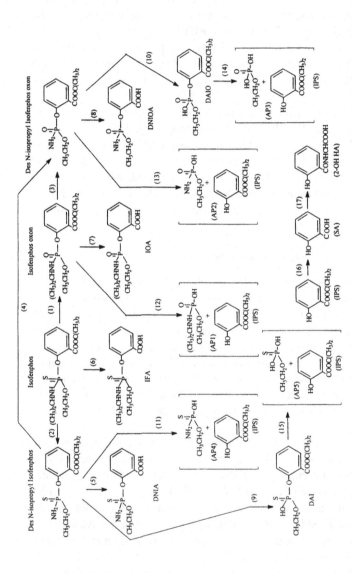

Fig. 3. Proposed metabolic pathway for isofenphos in the rat, guinea pig, and dog (from Knaak et al. 1996). (*1*) IF, isofenphos to isofenphos oxon; (*2*) IF, isofenphos to des N-isopropyl isofenphos; (*3*) isofenphos oxon to des N-isopropyl isofenphos oxon; (*4*) des N-isopropyl isofenphos to des N-isopropyl isofenphos oxon; (*5*) des N-isopropyl isofenphos to des N-isopropyl isofenphos oxon acid; (*6*) IF, isofenphos to isofenphos acid; (*7*) isofenphos oxon to isofenphos oxon acid; (*8*) des N-isopropyl isofenphos oxon to des N-isopropyl isofenphos oxon acid; (*9*) des N-isopropyl isofenphos to des-amino isofenphos; (*10*) des N-isopropyl isofenphos oxon to des-amino isofenphos oxon; (*11*) des N-isopropyl isofenphos to O-ethylaminothiophosphate and isopropyl salicylate; (*12*) isofenphos oxon to O-ethyliisopropylaminophosphate and isopropyl salicylate; (*13*) des N-isopropyl isofenphos oxon to O-ethyliisopropylaminophosphate and isopropyl salicylate; (*14*) des amino isofenphos oxon to O-ethylphosphate and isopropyl salicylate; (*15*) des amino isofenphos to O-ethylthiophosphate and isopropyl salicylate; (*16*) isopropyl salicylate to salicylic acid; (*17*) salicylic acid to 2-OH hippuric acid.

lyzed to give *O*-ethylphosphate (AP3) and *O*-ethylthiophosphate (AP5) (reactions 14 and 15). Approximately 75% of the isofenphos administered to the rat was eliminated in urine as alkyl phosphates and 18% as carboxylic acids of isofenphos, des *N*-isopropyl isofenphos or isofenphos oxon, and des *N*-isopropyl isofenphos oxon. In the case of the guinea pig and dog, 50% of the administered isofenphos was eliminated in urine as carboxylic acids and the remainder (~50% or less) as alkyl phosphates of isofenphos oxon, des *N*-isopropyl isofenphos oxon, and des *N*-isopropyl isofenphos. Isofenphos is less toxic to the guinea pig (LD_{50}, 600 mg/kg) than it is to the rat (LD_{50}, 40 mg/kg) (Wilson et al. 1984). This difference in toxicity appears to result in part from hydrolysis of the isopropyl ester by tissue carboxylesterases.

2. *Parathion*. Figure 4 gives the metabolic pathway for parathion in the rat (Menzie 1966; Neal 1967a,b; Nakatsugawa and Dahm 1967; Sultatos and Gagliardi 1990, Zhang and Sultatos 1991). The P=S group of parathion, like that of isofenphos, is converted to P=O (desulfurization) by P-450 isozymes to give paraoxon, or parathion is oxidatively cleaved to yield its respective alkyl phosphate (*O,O*-diethyl phosphorothioate) (reactions 1 and 2). The newly formed paraoxon may be directly hydrolyzed by paraoxonase to *O,O*-diethyl phosphate and *p*-nitrophenol (reaction 3). Free *p*-nitrophenol is conjugated in the liver to form *p*-nitrophenyl sulfate and glucuronide (reactions 4 and 5) or excreted per se.

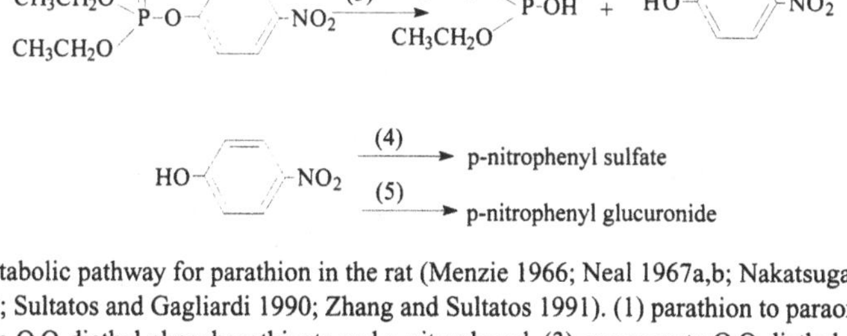

Fig. 4. Metabolic pathway for parathion in the rat (Menzie 1966; Neal 1967a,b; Nakatsugawa and Dahn 1967; Sultatos and Gagliardi 1990; Zhang and Sultatos 1991). (1) parathion to paraoxon, (2) parathion to O,O-diethyl phosphorothioate and p-nitrophenol, (3) paraoxon to O,O-diethyl phosphate and p-nitrophenol, (4) p-nitrophenol to p-nitrophenyl sulfate and (5) p-nitrophenol to p-nitrophenyl glucuronide.

3. *Malathion*. Figure 5 gives the metabolic pathway for malathion in the rat (Menzie 1966; Dauterman 1971). The major metabolites, malathion monoacid and diacid, were formed in the rat by the direct action of carboxylesterases (Knaak and O'Brien 1960) (products 5, 9 and 12). Malaoxon and the alkyl phos-

H.N. Nigg and J.B. Knaak

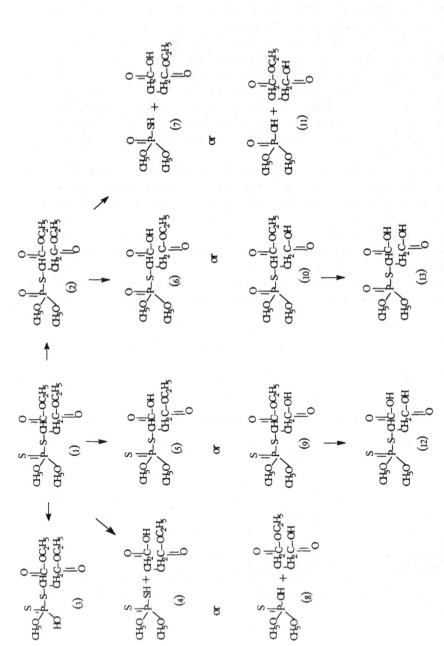

Fig. 5. Metabolic pathway for malathion. (1) malathion; (2) malaoxon; (3) desmethylmalathion; (4) O,O dimethyl-dithoic acid and succinic acid; (5 and 9) malathion α or β monoacid; (6 and 10) malaoxon α or β monoacid; (7) O,O-dimethyl phosphorothiolate and succinic acid; (8) O,O-dimethyl phosphorothionate and succinic acid; (11) O,O-dimethyl phosphoric acid and succinic acid; (12) malathion diacid; (13) malaoxon diacid.

phates, *O,O*-dimethyl phosphate and *O,O*-dimethyl phosphorothiolate, may be formed by the oxidative action of P-450 isozymes (products 2, 7 and 11) on malathion. However, phosphorotriesterases were implicated by Knaak and O'Brien (1960) in the formation of these alkyl phosphates and two other alkyl phosphates, *O,O*-dimethyl phosphorodithioate and *O,O*-dimethyl phosphorothionate, by their direct action on malathion (products 4 and 8). In practice, the dialkyl phosphorothiolates and phosphorothionates are not easily separated by ion-exchange chromatography and are considered to be equivalent. Small amounts of des methyl malathion were excreted by the rat (product 3). Two unknown products were present in urine; these products may be malaoxon mono- and diacid (products 6, 10 and 13).

IV. The Cholinesterases

The OP pesticides exert their insecticidal action by their ability to inhibit AChE at nerve synapses. The AChEs are high molecular weight proteins found in many tissues and particularly in nervous tissue per se catalyzing the hydrolysis of ACh after it is released at nerve synapses bringing about nerve impulse transmission. Several articles on this subject have been published (Aldridge 1954a,b; Aldridge and Davison 1953; O'Brien 1960).

A number of ChEs exist in the tissues of animals (i.e., nerve, red blood cells, serum, liver, etc.) and in insects capable of hydrolyzing choline esters (acetylcholinesterase, acetylcholine hydrolase [AChE], EC 3.1.1.7; butyrylcholinesterase [BChE], acylcholine acyl hydrolase, EC 3.1.1.8). The preferred substrate for AChEs is ACh, while BChEs prefer to hydrolyze butyrylcholine (BCh) and propionylcholine. AChEs and BChEs are classified as B-esterases, enzymes that are inhibited by OPs. Neuropathy target esterase (NTE), an enzyme involved in organophosphate-induced delayed neuropathy (OPIDN), is also classified as a B-esterase.

In humans, plasma BChE is depressed more quickly than RBC AChE and recovers more quickly after OP exposure ceases (Grob 1956; Grob and Harvey 1949, 1958; Grob et al. 1950, Grob et al. 1947a,b,c; Harvey et al. 1947). The ability of plasma BChE to be reactivated by pralidoxime is less in the fetus and pregnant women compared to nonpregnant women (Bell et al. 1979). Males have higher plasma BChE than females, and plasma BChE is more variable than RBC AChE (Augustinsson 1955). The active site serine is at amino acid 200 for RBC AChE and at amino acid 198 for plasma BChE (Lockridge et al. 1987b; Sutton et al. 1991). The source of RBC AChE is the bone marrow while BChE comes from the liver (Lockridge 1990). The physiological functions of RBC AChE and plasma BChE are not known (Brock and Brock 1993).

No genetic variants are known for human RBC AChE, but a 30% familial reduction has been reported (Johns 1962) and a reduction of 50% or more can occur with paroxysmal nocturnal hemoglobinuria (Kunstling and Rosse 1969). About 10 genetic variants are known for plasma BChE (Lockridge 1990) (Table 3).

Table 3. Human plasma cholinesterase phenotypes

Phenotype	Frequency of homozygote	Characteristics
Normal activity		
Usual	96:100	Normal activity
Reduced activity		
Atypical	1:3500	Dibucaine resistant
Silent	1:100,000	Very low activity, <2%
Fluoride	1:150,000	NaF resistant
Quantitative Var. J.	1:150,000	66% reduction
Var. K	1:100,000	33% reduction
Var. H	2 families	
Newfoundland	1 family	
Increased activity		
Cynthiana variant	4 families	2-3x normal
Johannesburg	1 family	2x normal

[a]From review by La Du et al. (1991) and Lockridge (1990). Also see Eckerson et al. (1983), Kalow and Genest (1957), Kalow and Gunn (1959), Kalow and Staron (1957), Krause et al. (1988), Neitlich (1966), Rubinstein et al. (1970), and Yoshida and Motulsky (1969).

A. Multiple Forms

AChE exists as mono-, di-, and tetramers of catalytic subunits (G_1, G_2, G_4) with each unit containing active sites. The most complex form, A_{12}, has 12 subunits. The forms are either readily extractable or tightly bound to membranes. The forms can be classified as either hydrophilic, water soluble or linked to a phospholipid membrane. The molecular differences in AChE and BChE were reviewed by Chatonnet and Lockridge (1989) and are listed in Table 4.

Ninety-five percent of human plasma BChE is the water-soluble G-4 form with no glycolipid anchor and no collagen tail. It is a tetramer of four identical subunits with a molecular weight of 340,000 and four active sites per molecule (Lockridge et al. 1979). The subunits are connected by two disulfide bonds (Lockridge et al. 1987a). The tetramer remains when the disulfide bonds are removed by proteolysis, reduction, or alkylation. Four active subunits can be generated with proteases or sonication. About 5% of human plasma BChE consists of monomer and dimers. Human plasma BChE appears to belong to a distinct family of serine esterases. The gene for human RBC AChE may be less than 60% identical with the gene for human plasma BChE. Enzymes with the same enzymatic properties as BChE have been isolated from human liver (Svensmark 1963), and liver transplantation changed a patient's plasma BChE from atypical-usual to homozygous-usual (Lockridge 1990).

In addition to separation by density gradient centrifugation, plasma and serum BChE were separated by starch or polyacrylamide gel electrophoresis (Bernsohn et al. 1961; Harris and Robson 1963; Hess et al. 1963; Juul 1968;

Table 4. Molecular polymorphism of acetylcholinesterase and butyrylcholinesterase.

Hydrophilic, water soluble forms secreted into body fluids

AChE			
G_1	G_2	G_4	A_{12}
Degradation product of G_4	Degradation product of G_4	Secreted by adrenal gland; peripheral nerve cells, found in plasma, spinal fluid	
BChE			
G_1	G_2	G_4	A_{12}
Degradation product of G_4	Degradation product of G_4	Human plasma BChE, 95% of activity	

Immobilized forms, asymmetrical

AChE			
G_1	G_2	G_4	A_{12}
			Torpedo californica muscle of primitive vertebrate
BChE			
G_1	G_2	G_4	A_{12}
			muscle of mammals and birds

Amphiphilic globular forms (membrane bound anchored to phospholipid bilayers)

AChE			
G_1	G_2	G_4	A_{12}
	Erythrocyte, glycolipid anchor-PI, cleaved by protease	Mammalian brain, 20-kDa anchor containing fatty acids	
BChE			
G_1	G_2	G_4	A_{12}
	Heart of *Torpedo marmorate* and superior cervical ganglion of rat	Detergent-soluble form reported in mammalian brain	

Source: Chatonnet and Lockridge (1989).
Globular forms: G_1, G_2, and G_4 contain one, two, or four subunits.
G_4 tetramer is an association of two dimers linked by disulfide bonds; dimers and monomers appear to be degradation products of tetramers. A_{12} contains three tetramers.

LaMotta et al. 1965, 1968; LaMotta and Woronick 1971; Lockridge and LaDu 1978; Kurono et al. 1979, 1983, 1984, 1991; Matsuzaki et al. 1980a,b; Yoshida et al. 1985; Masson 1979, 1989, 1991). Bernsohn et al. (1961) obtained at least six bands, Harris and Robson (1963) four components differing only in molecular size, Hess et al. (1963) obtained 6–7 bands, and Juul (1968) obtained 12 bands. LaMotta et al. (1965), using butyrylthiocholine as substrate, obtained 5 bands (C_1–C_5) of activity on starch gel. These bands all migrated to the C_5 position when they were eluted from an unstained gel, concentrated with ammonium sulfate, resolubilized, and separated by electrophoresis. When the 5 forms were eluted from unstained gel and individually subjected to electrophoresis without being treated with ammonium sulfate or being concentrated, the 5 bands interconverted, that is, band 1 converted to 3, 4, and 5 and so forth. About 96% of the plasma BChE was the C_5 form. LaMotta and Woronick (1971) presented evidence that plasma BChE enzymes C_1, C_2, C_3, C_4, and C_5 are isozymes. It is generally accepted that the usual plasma BChE consists of 4 bands on electrophoresis, C_1, C_2, C_3 and C_4. C_4 is a tetramer, C_2 a monomer conjugated to albumin, C_3 a dimer, and C_1 a monomer (Masson 1989, 1991). However, there is a fifth band in serum (Masson 1989) and up to 7 bands on starch gel electrophoresis (LaMotta et al. 1968).

Plasma BChE may be stored at $-70°C$ and frozen and thawed (10 times) with little change in activity or isozyme pattern. At $4°C$, the isozyme pattern changes were small after 30 d. At $22°C$, large changes were evident after 30 d (Juul 1968). Plasma BChE may be stored for several years at $-20°C$ without loss of activity, and outdated blood from blood banks is recommended for large studies because of the stability of plasma BChE (Whittaker 1986).

B. Three-Dimensional Structure, Amino Acid Sequence, and Active Sites

The three-dimensional structure of *Torpedo californica* AChE was determined by x-ray analysis after the dimer was solubilized by bacterial phosphatidylinositol (PI) -specific phospholipase C, purified by affinity chromatography, and crystallized from polyethylene glycol 200 (Sussman et al. 1991). The molecule has an ellipsoidal shape with dimensions ~45 Å by 60 Å by 65 Å. The monomer is an α/β protein that contains 537 amino acids and structurally is a 12-stranded mixed β sheet surrounded by 14 α-helices.

Amino acid sequences are known only for *Torpedo* AChE, *Drosophila* AChE, 85% of fetal bovine AChE, and human BChE. AChE and BChE have not been sequenced from the same species. The enzymes, however, have a high degree of similarity despite the fact that they are from different species (Chatonnet and Lockridge 1989). The active site serine is at amino acid 200 for RBC AChE and at amino acid 198 for plasma BChE (Lockridge et al. 1987b; Sutton et al. 1991).

The postulated "anionic site" of AChE that binds the quaternary ammonium ion of ACh is located in a gorge (Sussman et al. 1991). This site appears to be represented by 14 aromatic residues that line the gorge. Charges in the anionic site and active site are believed to stabilize the choline group. The active or esteratic site is embedded in a gorge about 20 Å long that reaches halfway into the three-dimensional structure of the protein and is responsible for the hydrolysis of ACh. An electrostatic attraction between the positive charge on the quaternary nitrogen atom of ACh and the negative change on the so-called anionic site on the enzyme forms the anionic substrate-binding site. A basic imidazole moiety (histidine) and an acidic moiety (tyrosine hydroxyl) at the active site catalyze the acetylation of a serine hydroxyl, followed by a rapid deacetylation restoring the enzyme and cleaving ACh into acetate and choline. There is evidence that the anionic site is uncharged and lipophilic (Sussman et al. 1991). The use of chiral OP compounds for AChE inhibition indicated two kinetically important regions separate from the esteratic site within 5 Å of the active site serine (Berman and Decker 1989; Berman and Leonard 1989).

Plasma BChE differs from RBC AChE in substrate specificity, pH optimum (Alles and Hawes 1940), substrate inhibition (Augustinsson 1948, 1949, 1960; Glick 1937; Wilson 1960; Zeller and Bissegger 1943), NaCl activation (Alles and Hawes 1940), amino acid sequence (Lockridge and LaDu 1978; Sutton et al. 1991), and organ source (Augustinsson 1960; Wilson 1960; Witter 1963); and the two enzymes are encoded by two distinct genes (Gnatt et al. 1991). Although AChE is produced from one gene, its properties are controlled by its tissue environment. Multiple mRNAs may lead to AChE structural diversity (Schumacher et al. 1988; Li et al. 1991, 1993a,b).

C. Distribution, Concentration, and Specific Activity

The distribution of AChE and BChE in tissues is given in Table 5 along with the concentration of these enzymes in blood plasma. The rate of hydrolysis (specific activity, 100–130 mmol of ACh hydrolyzed hr^{-1} mg^{-1} of protein) of ACh by

Table 5. Distribution of AChE and BChE in tissues.

Tissue	AChE	BChE
Muscle	Neuromuscular junction	Neuromuscular junction
Nervous system	Cholinergic system and in areas outside	In regions not related to AChE such as capillary endothelial cells, glial cells and neurones
Blood	54% plasma AChE in G4 form and 44% G1 and G2, 8 ng of AChE/mL in human plasma, AChE on red cell membranes	94% plasma BChE in water-soluble G4 form, origin the liver, 3300 ng of BChE/mL in human plasma
Heart	Nervous tissue	Nervous tissue

Source: Chatonnet and Lockridge (1989).

Table 6. Total activity of AChE and BChE in tissues.

Tissue	AChE Activity in nmol min^{-1} mL^{-1}	BChE Activity in nmol min^{-1} mL^{-1}
Plasma (human)	12.4 ±4.9	4219 ±834
Lumbar CFS (human)	13.3 ± 3.3	7.9 ±4.8
Ventricular CSF (human)	24.6 ± 1.7	34.9 ±15.6

CSF, Cerebrospinal fluid.
Activity determined using colorimetric method of Ellman et al. (1961) as modified by Bonham et al. (1981). CSF AChE activity determined using 0.5 x 10^{-3} M acetyl-β–methylthiocholine in the presence of 1.4 × 10^{-5} M ethopropazine (BChE inhibitor) plasma and CSF activities determined using 0.5 × 10^{-3} M butyrylcholine as substrate (no inhibitors).
Source: Atack et al. (1987).

purified electric eel AChE was measured by Schaffer et al. (1973) at pH 7.4, 25°C, using 0.0037 M ACh. The activity of AChE and BChE in 1.0 mL of several body fluids is given in Table 6.

V. Acylation and Phosphorylation of Cholinesterase and Binding of Substrates and Organophosphorus Inhibitors to Cholinesterase

The reaction of ChEs with substrates and their inhibition by OP insecticides (inhibitors) are considered to occur by an analogous mechanism (Aldridge 1954a,b; Wilson et al. 1950; Wilson and Bergmann 1950; Wilson 1960). ChEs rapidly hydrolyze esters of choline, and esters of phosphorus (OPs), slowly. The first product, choline, is liberated leaving an acetyl group bound to the enzyme. The acetyl group is then rapidly cleaved from the enzyme, resulting in the formation of acetic acid (Fig. 6), whereas in the case of phosphate ester inhibitors, the phosphoryl group remains bound to enzyme preventing the enzyme from reacting with another molecule, substrate, or inhibitor. Very good inhibitors are found to be compounds that resemble the structure of substrates. OP compounds that resemble ACh in this respect are potent inhibitors of ChEs; for example,

$$CH_3 \diagdown \overset{O}{\underset{F \diagup}{\overset{||}{P}}}-O(CH_2)_2N^+CH_3$$

which has a rate constant of inhibition (phosphorylation) of AChE (2.3×10^8 M^{-1} min^{-1}), approaching the rate constant of acetylation by ACh (>3×10^9 M^{-1} min^{-1}) (Tammelin 1958a,b,c).

The reaction of OPs generally requires activation to the oxon analog but not always (Aldridge 1954b; Chambers 1992; Wallace 1992). Some pesticides, e.g., acephate and methamidophos, are oxon OPs and require no activation. Simply stated, activation is the replacement of the double-bonded sulfur on the phospho-

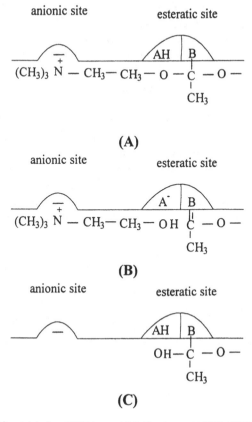

Fig. 6. Hydrolysis of acetylcholine (ACh) by acetylcholinesterase (AChE). (A) Binding of positively charged quaternary nitrogen to anionic site and ester to esteratic site. (B) Hydrolysis of ACh to liberate choline. (C) Release of acetate. (Figure redrawn from O'Brien 1960.)

rous atom with double-bonded oxygen (Wallace 1992; see Chambers 1992 for a review of OP pesticide activation).

The original description of the inhibition process by Hart and O'Brien (1973) was a two-step process: equilibrium (occupation) of the OP compound with the active site of AChE followed by covalent bonding (phosphorylation) of the phosphorus of the OP to the oxygen of the hydroxyl group of serine at the active site (Michel and Krop 1951; Wilson 1951; Aldridge and Davison 1953; Oosterbaan et al. 1955). Beyond the covalent bond formed by OP compounds with the active site serine, the description of the inhibition of AChE by OPs and other inhibitors becomes increasingly complicated. The OP compounds undergo a process termed 'aging' after covalent bond formation. For most OP pesticides, an ethoxy or methoxy group is cleared from the portion of the OP molecule covalently bound to the serine hydroxyl group. Once aging occurs, the enzyme is dead. New enzyme must be synthesized. The aging process is the reason that RBC AChE activity returns at about 1%/d, the approximate rate of the synthesis of RBCs in humans. For plasma BChE, the return of activity is higher, up to 16%/ d. For a more complete (and complex) explanation of the OP–AChE inhibition process, we refer the reader to Berman (1995) and to Soreq and Zakut (1993).

There are other possibilities for inhibition of AChE and BChE. Fascisulin, a green mamba 61-amino-acid peptide, appears to be a peripheral site inhibitor of AChE. Fascisulin inhibits AChE approximately 1 million times better than it inhibits BChE (Radic et al. 1995). Sulfonyl fluro dyes are irreversible inhibitors of ChEs and form a covalent sulfonyl–enzyme complex similar to the OP compounds (Moss et al. 1988). In monkey experiments with methanesulfonyl fluoride, RBC AChE was reduced to zero with no symptoms. Three days later cerebral cortex AChE was 80% inhibited (Moss et al. 1988). Recognizing that differences in AChE can occur in different tissues and also in the same tissue, that an inhibitor can be selective for AChE or BChE, and that blood esterases can be reduced to zero without symptoms, can we predict, with safety, the inhibition of central nervous system esterases by monitoring blood esterases?

A. Hydrolysis of Acetylcholine by Acetylcholinesterase

In general terms, AChE (EH) combines with substrate (AB) (i.e., ACh) resulting in an enzyme–substrate complex (Michaelis complex) (EHAB) that is converted to an acylated enzyme (EA) and a leaving group (BH) (Aldridge and Reiner 1972). The acylated enzyme then decomposes into an enzyme product as indicated below:

$$\overset{k_{+1}}{\underset{k_{-1}}{EH + AB \rightleftharpoons EHAB}} \overset{k_{+2}}{\rightarrow} EA + BH \overset{k_{+3}}{\underset{{}^+H_2O}{\rightarrow}} EH + AOH \tag{1}$$

where k_{+1}, k_{-1}, k_{+2} and k_{+3} are the velocity (in min^{-1} or hr^{-1}) of the equilibrium reactions. BH and AOH are the products of substrate hydrolysis; if AB is ACh, BH is choline and AOH is acetic acid, respectively. The acetylated enzyme (EA) formed from ACh (Wilson 1951) hydrolyzes very quickly (half-life, 2.3×10^{-6} min or less) (Cohen et al. 1955)).

The rate of product formation is described by the following equation:

$$v = k_{+3}[EH] \tag{2}$$

When the reaction described by Eq. 1 is at steady state, the rates of formation of EHAB and EA will equal the rates of their breakdown and the following two equations will hold:

$$k_{+1}[EH][AB] = (k_{-1} + k_{+2})[EHAB] \tag{3}$$

$$k_{+2}[EHAB] = k_{+3}[EA] \tag{4}$$

The total enzyme concentration $[E_t]$ is equal to [EH], the concentration of the free enzyme, plus [EHAB], the concentration of the enzyme combined with substrate or the concentration of acylated enzymes, [EA]:

$$[E_t] = [EH] + [EHAB] + [EA] \tag{5}$$

Inserting Eq. (4) into Eq. (3) gives:

$$[EH] = \frac{k_{+3}(k_{-1} + k_{+2})[EA]}{k_{+1}k_{+2}[AB]} \quad (6)$$

and inserting Eq. 6 and 4 into Eq. 5 gives:

$$[Et] = \frac{k_{+3}(k_{-1} + k_{+2})[EA]}{k_{+1}k_{+2}[AB]} + \frac{k_{+3}[EA]}{k_{+2}} + [EA] \quad (7)$$

or after rearrangement:

$$[E_t] = [EA] \frac{k_{+3}(k_{-1} + k_{+2}) + k_{+1}[AB](k_{+2} + k_{+3})}{k_{+1}k_{+2}[AB]} \quad (8)$$

Inserting Eq. 8 into Eq. 2 gives:

$$v = \frac{k_{+1}k_{+2}k_{+3}[Et][AB]}{k_{+3}(k_{-1} + k_{+2}) + k_{+1}[AB](k_{+2} + k_{+3})} \quad (9)$$

or rearranged:

$$v = \frac{(k_{+2} + k_{+3})/(k_{+2} + k_{+3})[Et][AB]}{k_{+3}(k_{-1} + k_{+2})/k_{+1}(k_{+2} + k_{+3}) + [AB]} \quad (10)$$

$$V_{max} = \frac{(k_{+2} + k_{+3})}{(k_{+2} + k_{+3})}[Et] \quad (11)$$

$$K_m = \frac{k_{+3}(k_{-1} + k_{+2})}{k_{+1}(k_{+2} + k_{+3})} \quad (12)$$

$$v = \frac{V_{max}[AB]}{K_m + [AB]} \quad (13)$$

Equation 13 is the Michaelis equation for Eq. (reaction) 1. The Michaelis equation relates enzyme activity (v) with the substrate concentration under steady-state conditions when the concentrations of all enzyme species (EH, EHAB, and EA) are constant. K_m and V_{max} are the Michaelis constant and the maximum rate of substrate hydrolysis.

The formation of the Michaelis complex (k_{+1} in Eq. 1) is the fastest of all reactions. Decomposition of the Michaelis complex is fast (k_{+2} in Eq. 1), and it is only within the last few years that methods have become available to measure its rate. V_{max} and K_m may be determined graphically from experimental data by plotting $1/v$ versus $1/(S)$. The line intercepts the Y-axis at $1/V_{max}$, while the slope of the line is K_m/V_{max}.

The catalytic center activity (CCA) as defined in Equation 14 is the number of ester molecules hydrolyzed by one active center in 1 min. Substrates react quickly, and most studies are therefore performed under steady-state conditions, because steady state is reached quickly (Aldridge and Reiner 1972).

$$\frac{V_{max}}{[E_t]} = \frac{k_{+2}k_{+3}}{k_{+2} + k_{+3}} \quad (14)$$

B. Phosphorylation of Cholinesterase by Organophosphorus Inhibitors

The irreversible inhibition of AChE is studied in vitro by the direct addition of several inhibitor concentrations (OPs) to AChE in aqueous medium in the absence of substrate or by the addition of inhibitor concentrations to AChE in aqueous medium in the presence of substrate. In the absence of substrate, OPs rapidly inhibit the activity of AChE. Inhibition is stopped and residual activity is determined by adding ACh and measuring the rate of hydrolysis (i.e., liberation of acetic acid) using a pH stat technique. In the presence of substrate, OPs react slowly with AChE and progressive inhibition is followed by determining the rate of hydrolysis of ACh through time. Steady state is reached slowly and the reactions can be studied under pre-steady-state conditions.

1. *Rate of Phosphorylation (Inhibition) in the Absence of Substrate.* Main (1964) studied the irreversible inhibition of AChE by OPs as represented by Eq. 15:

$$\underline{EH} + \underline{AB} \underset{k_{-1}}{\overset{k_{+1}}{\rightleftharpoons}} EHAB \overset{k_{+2(p)}}{\rightarrow} \underline{EA} + BH \tag{15}$$

where EH is AChE (EC type 3.1.1.7 hydrolase), AB is an OP inhibitor, EHAB is a reversible complex, and EA is the irreversibly phosphorylated enzyme. The individual rate constants are k_{+1}, k_{-1} and $k_{+2(p)}$. The equilibrium constant or affinity constant, K_a, is similar to K_m in Eq. 12 and is given by

$$K_a = \frac{k_{-1}}{k_{+1}} \tag{16}$$

The phosphorylation constant, $k_{+2(p)}$, gives the rate of formation of the irreversible phosphorylated enzyme and the release of the leaving group BH. A bimolecular rate constant, k_i, includes the constants for affinity and phosphorylation, and $k_i = k_{+2(p)} / K_a$.

Equation 17 was derived from Eq. (reactions) 15 and includes K_a (Main 1964) by taking into account the initial concentrations of EH (e) and AB (i) and the concentrations of EHAB (r) and EA (q). The reaction between (e) and (i) follows first-order kinetics over significant ranges of inhibitor and enzyme concentrations when (i) is constant. The fraction of the esterase in the reversible form EHAB then bears a constant relationship to the total available enzyme. The ratio [$r/(e - q)$] must be constant. The reversible reaction approximates steady state when

$$k_{+1}(e - r - q)i = k_{-1}r \tag{17}$$

and

$$r = (e - q)i/i + K_a \tag{18}$$

The rate of irreversible inhibition is

$$dq/dr = k_{+2\,(p)}\,r \tag{19}$$

Substituting Eq. 18 into Eq. 19 gives

$$dq/dt = i/(i + K_a)\,k_{+2(p)}\,(e - q) \tag{20}$$

Integrating Eq. 20 between the limits of q_1 and q_2 and t_1 and t_2 observing that log $(e - q_1)/(e - q_2) = \log(v_1/v_2) = (\Delta \log v)$ for the interval (Δt) gives

$$\frac{1}{i} = \frac{\Delta t}{2 \cdot 3 \Delta \log v}\left(\frac{k_p}{K_a} - \frac{1}{K_a}\right) \tag{21}$$

Substituting Eq. 22 into Eq. 21 gives Eq. 23:

$$k_i = k_p/K_a \tag{22}$$

$$\frac{1}{i} = \frac{\Delta t}{2 \cdot 3 \Delta \log v}\left(k_i - \frac{1}{K_a}\right) \tag{23}$$

Main (1964) plotted the reciprocal of the inhibition rate $(\Delta t/2.3\,\Delta \log v)$ against the reciprocal of the inhibitor concentration $(1/i)$ according to Eq. 23 for the reactions of DFP and malaoxon with human serum ChE at 37°C, pH 7.6. The slopes gave k_i, the bimolecular rate constant, and the intercepts of the extrapolated lines on the ordinate and abscissa gave $(-1/K_a)$ and $(1/k_p)$. The K_a and k_p values for malaoxon were 7.7×10^{-4} M and 11 min^{-1}, respectively. A value for k_i could be calculated for each point on the line by rearranging Eq. 23 to Eq. 24.

$$k_i = \frac{k_p}{k_a} = \frac{2 \cdot 3 \Delta \log v}{\Delta t}\left(\frac{1}{i} + \frac{1}{K_a}\right) \tag{24}$$

The principle of the procedure is the same as that employed by Aldridge (1950), in which esterase and an excess of inhibitor was incubated for a measured time (in minutes) after which the reaction was stopped by the addition of substrate and the residual activity measured. The substrate reactions were later followed by Main and Iverson (1966) in special reaction vessels, jacketed for temperature control and with an interior capacity of 75 mL. Incubation times were of the order of 1–10 sec. The inhibition reaction was stopped by rapidly adding substrate to the reaction. Acid liberated by hydrolysis of substrate was determined using a pH stat. The conditions were changed because the determination of K_a and k_p for DFP required much higher inhibitor concentrations and shorter incubation periods than were required for malaoxon.

2. *Rate of Phosphorylation (Inhibition) in the Presence of Substrate.* OP compounds in the presence of substrate react with esterases to produce relatively stable phosphorylated enzymes. The reactions according to Main (1973) may be described by the following two equations:

$$\overset{K_a}{}\qquad\overset{k_2}{}$$
$$\text{EH} + \text{AX} \rightleftharpoons \text{EHAX} \rightarrow \text{EA} + \text{HX} \tag{25}$$
$$\underset{k_i}{\overline{}}$$

$$\text{EH} + \text{CB} \underset{k_{-1}}{\overset{k_{+1}}{\rightleftharpoons}} \text{EHCB} \overset{k_2}{\searrow} \text{EC} + \text{HOH} \overset{k_3}{\rightarrow} \text{EH} + \text{AOH} \tag{26}$$
$$\text{HB}$$

where EH is free enzyme, AX is an ester of an OP acid, EHAX is the enzyme OP complex, EA is the phosphorylated enzyme, and HX is the first leaving product; CB is the substrate, EHCB is the enzyme–substrate complex, EC is the acetyl enzyme, HB the leaving group, and EH the regenerated enzyme. K_a is the dissociation or affinity constant for the enzyme–inhibitor complex. Main (1973) worked out a procedure for evaluating K_a, k_{+2}, and k_i. In conditions where enzyme<<inhibitor and enzyme<<substrate, the enzyme catalyzing the substrate is progressively removed by the action of the OP inhibitor present in the substrate, causing the reaction curve to deviate from linearity. The curve is analyzed by drawing tangents (slopes = v) at time t. A plot of log v against t gives the first-order inhibition rate in min^{-1}, where the slope of the line is

$$p = \frac{2 \cdot 3 \log(V_o/V)}{t} = \frac{k_{+2}}{1 + (K_a/i)} \tag{27}$$

To determine K_a and k_{+2}, a set of four or five log v versus t plots are obtained using an appropriate range of inhibitor. The slopes, p, and inhibitor concentrations are used to construct plots of $1/p$ versus $1/[I]$. A plot of $1/p$ versus $1/[I]$ gives the values for $1/k_i$ (slope), $1/k_{+2}$ (y-intercept), and $-1/K_a$ (x-intercept) according to the following equation:

$$1/p = \frac{K_a}{k_{+2}} \times \frac{1}{i} + \frac{1}{k_{+2}} \tag{28}$$

The bimolecular rate constant, k_i, is related to K_a and k_{+2} by

$$k_i = k_{+2}/K_a \tag{29}$$

Hart and O'Brien (1973) developed another procedure based on the use of enzyme, inhibitor, and chromogenic substrate in a buffer. This system is depicted below:

$$\text{E} + \text{PX} + \text{S} \underset{\substack{k_{-1} \\ k_{+1}' \\ k_{-1}'}}{\overset{k_{+1}}{\rightleftarrows}} \begin{array}{l} \text{EPX} \overset{k_{+2}}{\longrightarrow} \text{EP} + \text{X} \\ \text{ES} \overset{k_{+2}'}{\longrightarrow} \text{E} + \text{products} \end{array} \tag{30}$$

Experimental conditions were selected so that the molar concentrations of inhibitor and substrate are each greater (>10^5) than that of enzyme-active sites.

Under these conditions, the enzyme–substrate and enzyme–inhibitor reactions are both pseudo-first order and the rate of reaction along either pathway is proportional to the concentration of nonirreversibly inhibited enzyme. The chromophore, p-nitrophenol, is generated at a rate directly related to the rate of phosphorylation of the enzyme. A reaction involving paraoxon (60 units of AChE mL^{-1}, 2 x 10^{-5} M paraoxon, and 2 mM p-nitrophenyl acetate at 25°C) gave an absorbance–time curve starting at zero and curving upward. The reaction velocities (slopes of tangents to the curve taken at various times) plotted against time were used to obtain the zero-time velocity for each inhibitor concentration. The following equation was used to allow for the effect of substrate, where v_0 is the zero-time velocity at a given concentration of inhibitor, K_s is the dissociation constant of the ES complex, and V_m is the maximum velocity for the enzyme–substrate reaction in the absence of inhibitor:

$$v_0 = \frac{V_m[S]}{K_s[1+([PX]/K_d)]+[S]} \tag{31}$$

Introducing $V_m = v[1 + (K_m / [S])$ (Briggs–Haldane corrected Michaelis–Menten equation), the equation becomes

$$\frac{v_0}{v} = \frac{K_m+[S]}{K_m[1+([PX]/K_d)]+[S]} \tag{32}$$

K_d was determined using K_m as the nearest available approximation of K_s, by rearrangement of the equation to give

$$K_d = \frac{K_m+[PX]}{(K_m+[S])(v_c/v_0-1)} \tag{33}$$

where v_c is the velocity of a control reaction carried out in the absence of inhibitor, but at the same [S] used in the inhibition reaction.

The phosphorylation constant (k_{+2}) in Eq. 35 was derived by considering the relationship between the various forms of the enzyme as shown in Eq. 34 according to the procedure of Main (1964) except for the fact that allowance was made for the effect of substrate.

$$[Ef] = [E]_0 - [EP] - [EPX] - [ES] \tag{34}$$

where [Ef] = free enzyme, [E]$_0$ = initial enzyme, [EP] = phosphorylated enzyme, [EPX] = enzyme–OP complex, and [ES] = enzyme–substrate complex.

$$k_{+2} = \frac{\Delta \ln v}{\Delta t}\left(\frac{K_d}{[PX](1-\alpha)}+1\right) \tag{35}$$

where $\alpha = [S]/(K_m + [S])$

The phosphorylation rate constant (k_{+2}) was obtained by inserting the value for K_d determined by the zero-time method (Eq. 33).

Hart and O'Brien (1973), using the zero time method, obtained the following values for the inhibition of AChE by paraoxon at 25°C: $K_d = 0.55 \times 10^{-4}$M, $k_{+2} = 14.18$ min^{-1}, and $k_i = 25.88 \times 10^4$ (M^{-1} min^{-1}).

C. Inhibition of Multiple Forms of Acetylcholinesterase and Butyrylcholinesterase by Organophosphorus Inhibitors

Main (1969) showed that inhibition of serum ChE and RBC AChE by DFP and amiton (*O,O*-diethyl *S*-2-diethylamino ethylphosphorothiolate) did not follow first–order kinetics. The rate plots curved (concave) and were not linear with time. The curving was believed to be caused by multiple enzyme forms.

The inhibition of serum (horse and human) and RBC AChE (bovine) by DFP, amiton, and malaoxon was used by Main (1969) to determine the activity of multiple forms of serum ChE and RBC AChE using 30 mM BCh iodide and ACh chloride, respectively, as substrates. A solution of the horse serum (1.5 mg/mL) purified by the Strelitz method (Strelitz 1944) from Nutritional Biochemicals contained active site concentrations of 6.4×10^{-8} moles g^{-1} of dried powder while human serum fraction (IV-6-3 ChE-Type II from Sigma) contained active site concentrations of 2.35×10^{-7} moles g^{-1} of protein. La Motta et al. (1965) resolved serum fraction IV-6-3 into five or more active bands by starch gel electrophoresis.

The *in vitro* inhibition rates were determined by allowing the inhibitor to react for a measured time, after which the inhibition was stopped by the addition of substrate and the residual activity measured. The inhibition and substrate reactions were carried out at pH 7.0. A Radiometer pH-stat was used to titrate the liberated acetic acid during the course of the reaction. The curved rate plots (upward and concaved) obtained with partially purified horse ChE were resolved into four linear components at 5°C and three linear components at 25°C. Each resolved component represented the first-order rate plot for the inhibition of one form of ChE, indicating that four forms were present in the sample. The first-order rate constants (ρ) and the intercept values (v_0) for the multiple forms of human IV-6-3 ChE obtained by Main (1969) using regression analysis on log Σv versus t plots are given in Table 7. ChE I was the form most rapidly inhibited by amiton, followed by ChE II and so forth.

According to Main (1969), the time course of the inhibition of the individual enzymes (AChE and ChE) follows first-order kinetics when inhibitor concentrations [I] remain constant throughout a study. The first-order rate constants for individual enzymes were defined experimentally as $\rho = 2.3$ (log v_0 − log v)/t where (v_0) is the initial substrate concentration and $v \propto e$. The velocity measured experimentally after inhibition time (t) is the sum of the individual velocities of the multiple forms, $\Sigma v = v^I + v^{II} + v^{III} = v_0{}^I e^{-\rho It} + v_0{}^{II} e^{-\rho IIt} + v_0{}^{III} e^{-\rho IIIt}$ where e is the base of the natural logarithms and the properties of the individual enzymes are identified by the superscripts I, II, and III. The forms were numbered in order of decreasing rates of inhibition, the most rapidly inhibited was ChE I, the next ChE II, and so forth. Initially, all forms are active, followed by the complete inhibition

Table 7. First-order rate constants (ρ) and intercept values (v_0) of multiple forms of human IV-6-3 ChE inhibited by various concentrations of amiton at pH 7.0, 5°C.

i	ChE I		ChE II		ChE III	
	ρ^I	$v_0{}^I$	ρ^{II}	$v_0{}^{II}$	ρ^{III}	$v_0{}^{III}$
M						
1.07×10^{-6}	5.3 ± 0.3	233	1.1 ± 0.3	28.5	0.11 ± 0.001	10.5
8.0×10^{-6}	28.7 ± 2.0	237	2.7 ± 0.3	27.0	0.18 ± 0.01	10.7
1.33×10^{-5}	36.1 ± 3.1	205	3.4 ± 0.4	27.2	0.21 ± 0.02	11.1
2.0×10^{-5}	45.1 ± 2.4	233	2.9 ± 0.4	22.9	0.27 ± 0.02	10.1
2.67×10^{-5}	57 ± 4.0	246	3.0 ± 0.2	28.9	0.24 ± 0.03	7.64
Average		231 ± 15		26.9 ± 2.4		10.0 ± 1.8

The ρ and v_0 values were calculated by regression analysis. The final ChE concentration was 1.01×10^{-7} M in 3.3 mM sodium phosphate buffer. One velocity unit is 0.01 mmole per min at 25°C, pH 7.0; p is min^{-1} $\pm$ SD; i is inhibitor concentration.
Source: Main 1969.

of ChE I and ChE II. When ChE II is completely inhibited, $\Sigma v = v^{III}$, and in this terminal region the log Σv against t plot will be linear. For significant curving to occur, at least two of the (ρ) values should differ with (v_0) values being similar. In his studies, Main (1969) assumed that a reversible complex was formed before the phosphorylation of the active site.

The basic scheme was where E, I, EI, and E' are the enzyme, inhibitor, reversible complex, and

$$E + I \underset{k_{-1}}{\overset{k_1}{\rightleftharpoons}} EI \overset{k_2}{\rightarrow} E'$$

phosphorylated active site, respectively. When $k_{-1} \approx k_2$, the formation of (EI) is controlled by the equilibrium affinity constant, $K_a = k_{-1} / k_1$ and $\rho = k_2 / (1 + K_a/i)$.

When $k_{-1} \gg k_2$, the solution to the second-order differential yields an expression containing two exponential terms. One of these disappears when the transient phase of the reaction in which EI is increasing from 0 is over; then $\rho = [k_{-1} + k_2 + k_1 i - \sqrt{((k_{-1} + k_2 + k_1 i)^2 - 4k_1 k_2 i)}] / 2$ and is constant (Main 1969).

The inhibition of Strelitz horse ChE by amiton and DFP at 5°C revealed significant differences in the kinetic properties of the ChE forms (Main 1969). Differences between rates of phosphorylation (k_2) of successive forms varied from 8- to 50 fold whereas differences in binding (K_a) varied from 1.7- to 13 fold. Differences in inhibitory power (k_i) were intermediate between k_1 and K_a varying between 2.2- and 23 fold. The 43-fold difference between the first-order rate constants of ChE I and ChE II observed by Main (1969) when Strelitz horse ChE inhibited by amiton at 25°C was reduced to a calculated 1.5-fold difference with malaoxon as inhibitor.

The studies of Main (1969) strongly suggest that definitive *in vitro* studies are needed to obtain kinetic data on the inhibitory action of OPs on serum, RBC, or tissue ChEs before their effects can be adequately assessed in animal toxicity studies or the results extrapolated to humans.

VI. Kinetic Analysis of Species Organophosphorus Compound Sensitivity

The USEPA requires pesticide registrants to conduct animal toxicity studies, principally in the rat and mouse, for purposes of extrapolating the results in risk assessment scenarios to humans. According to Murphy (1972), different species vary widely in their sensitivities to OP pesticides. The differences often could not be explained on the basis of metabolism alone (i.e., rates of activation and detoxification) (Murphy 1966; Hitchcock and Murphy 1967). *In vitro* incubation of brain AChE from several species with OPs indicated that the concentrations required to produce 50% inhibition of AChE activity (IC_{50}) differed from species to species using the same OP (Murphy et al. 1968). The IC_{50} of gutoxon (oxon of azinphosmethyl) for chicken brain was 30 fold higher than that for rat brain and 43 fold higher for sunfish brain AChE. The differences were believed to be caused by differences in binding affinities (K_a) and rates of phosphorylation (k_p). Wang and Murphy (1982) developed IC_{50} values for DFP, methyl paraoxon, paraoxon,

Table 8. Concentration of organophosphate compound required to inhibit 50% acetylcholinesterase activity at 22°C in 30 min, pH 8.0.

Enzyme source	DFP (M)	Methylparaoxon (M)	Paraoxon (M)	Gutoxon (M)	Ethyl gutoxon (M)
Monkey	6.40×10^{-7}	6.75×10^{-8}	1.70×10^{-8}	1.22×10^{-7}	1.36×10^{-8}
Rat	2.02×10^{-6}	1.93×10^{-8}	1.17×10^{-7}	1.28×10^{-8}	1.36×10^{-9}
Guinea pig	5.30×10^{-6}	3.01×10^{-7}	1.07×10^{-7}	6.32×10^{-8}	2.96×10^{-8}
Chicken	9.70×10^{-7}	1.48×10^{-8}	6.20×10^{-9}	3.57×10^{-7}	7.75×10^{-8}
Catfish	2.24×10^{-6}	3.73×10^{-7}	1.80×10^{-7}	2.13×10^{-8}	3.51×10^{-8}
Frog	6.07×10^{-5}	1.53×10^{-6}	6.25×10^{-7}	1.36×10^{-6}	4.81×10^{-7}
Magnitude of difference[a]	94.8×	103.4×	100.8×	106.3×	68.4×

DFP, diisopropyl fluorophosphate.
[a]Magnitude of difference between most sensitive and least sensitive species.
Source: Wang and Murphy (1982).

gutoxon, and ethyl gutoxon using brain synaptosomal AChE from the monkey, rat, guinea pig, chicken, catfish, and frog. A difference of approximately 100 fold was observed for each OP between the IC_{50} value for the most sensitive and least sensitive species. The IC_{50} values are given in Table 8. Frog brain AChE was the least sensitive to all the OPs studied, with chicken brain being the most sensitive to methyl paraoxon and paraoxon. Rat brain AChE was the most sensitive to gutoxon and ethyl gutoxon. Monkey brain AChE was the most sensitive to DFP.

Wang and Murphy (1982) determined the affinity and phosphorylation constants (K_a and k_p) for each OP (DFP, paraoxon, methylparaoxon, gutoxon, and ethylgutoxon) and source of AChE (monkey, rat, guinea pig, chicken, catfish, and frog) using the method of Wustner and Fukuto (1974). This method is similar to the one described in Section V.B by Main (1964) (rate of phosphorylation in absence of substrate, use of pH stat) except that DTNB was added with substrate to colorimetrically measure the hydrolysis of acetylthiocholine. The method differs from that of Hart and O'Brien (1973) where inhibitor (paraoxon) and substrate (p-nitrophenylacetate) were added at the same time and hydrolysis of substrate (formation of p-nitrophenol) was followed by spectroscopy. The inhibition constants were calculated according to Eq. 24. The IC_{50}s were calculated from the bimolecular rate constant (k_i) based on the equation, $\log P = 2 - (k_i \times i / 2.303 \times t)$ (O'Brien 1960). The calculated pI_{50}s were compared to those obtained experimentally ($pI_{50s} = -\log [IC_{50}]$). Tables 9 and 10 give the results obtained with paraoxon and methylparaoxon.

Table 9. Affinity equilibrium (K_a), phosphorylation rate (k_p), and bimolecular rate (k_i) constants for the inhibition of brain acetylcholinesterase by paraoxon at 22°C, pH 8.0.

Enzyme source	K_a (μM)	k_p (min^{-1})	k_i (μM^{-1}min^{-1})	pI_{50} calculated	pI_{50} observed
Monkey	10.96 ± 0.45	13.42 ± 0.42	1.224	7.724	7.770
Rat	21.69 ± 4.72	38.17 ± 6.82	1.760	7.882	7.932
Guinea pig	198.68 ± 23.3	21.32 ± 1.77	0.107	6.666	6.970
Chicken	6.35 ± 0.58	25.26 ± 1.19	3.978	8.236	8.208
Catfish	202.7 ± 20.5	22.18 ± 1.58	0.109	6.674	6.745
Frog	90.85 ± 22.9	3.62 ± 0.26	0.040	6.237	6.204

Standard errors were calculated by the regression method of Wilkinson (1961).
Source: Wang and Murphy (1982).

The longer alkyl group of paraoxon (ethyl) compared to methyl paraoxon resulted in higher affinities (lower K_as) by a factor of more than 2.0 with the phosphorylation constants remaining approximately the same. The higher K_a values for methylparaoxon resulted in lower bimolecular rate constants (k_is) than were obtained with paraoxon and lower toxicity.

Table 10. Affinity equilibrium (K_a), phosphorylation rate (k_p), and bimolecular rate (k_i) constants for the inhibition of brain acetylcholinesterase by methylparaoxon at 22°C, pH 8.0.

Enzyme source	K_a (μM)	k_p (min^{-1})	k_i (μM^{-1}min^{-1})	pI$_{50}$ calculated	pI$_{50}$ observed
Monkey	26.79 ± 1.28	11.79 ± 0.37	0.440	7.280	7.171
Rat	39.48 ± 9.77	38.29 ± 8.47	0.970	7.623	7.714
Guinea Pig	459.92 ± 45.6	23.83 ± 1.80	0.052	6.352	6.521
Chicken	24.9 ± 2.12	37.86 ± 2.40	1.517	7.817	7.830
Catfish	241.7 ± 28.5	14.74 ± 1.07	0.061	6.422	6.428
Frog	177.4 ± 33.7	2.45 ± 0.16	0.014	5.777	5.815

Standard errors were calculated by the regression method of Wilkinson (1961).
Source: Wang and Murphy (1982).

The source of enzyme (i.e., monkey, rat, etc.) had an even greater effect on K_as than the length of the alkyl group. The difference in K_as suggests that the structure of AChE varies somewhat between species. No K_a values were obtained with human brain AChE. The k_p values also suggest differences in enzyme structure or in the nature of the phospholipid membranes to which they are attached. A similar study by Cohen et al. (1985) using brain homogenates from the cow, purified bovine RBC AChE, dyfonate oxon, paraoxon, and malaoxon showed larger K_a and k_p values with cow brain AChE than obtained with purified bovine RBC AChE. However, the bimolecular rate constants (k_i) were almost identical. The bovine brain AChE k_i values were 0.373, 0.398, and 0.130 μM^{-1} min^{-1} for dyfonate-oxon, paraoxon, and malaoxon, respectively. The k_i for rat brain AChE was 1.38 μM^{-1} min^{-1} in the study by Cohen et al. (1985) with paraoxon and 1.76 μM^{-1} min^{-1} in the study by Wang and Murphy (1982). The paraoxon rat brain affinity (K_a) and phosphorylation constants were both lower in the Cohen et al. (1985) study. The k_i (IC$_{50}$) values, therefore, are less affected by experimental conditions and should be used.

VII. Neurotoxicity: Effects on Behavior, Brain, and Blood AChE Activity

The pharmacological effects of OPs were believed to be of short duration, entirely caused by the inhibition of AChE and dependent upon the regeneration of AChE (Srinivasan et al. 1976). However, in a review by Karczmar (1984) the evidence presented suggests that anti-ChEs may directly affect a second messenger as well as a transmitter system other than the cholinergic system.

Goldberg et al. (1963, 1965) were the first to examine the relationship between the anti-ChEs (i.e., eserine, *N*-methyl 3-isopropyl phenyl carbamate [compound 10854] and carbaryl) and discrete avoidance behavior. Evidence presented by Goldberg et al. (1965) indicates that reduction in AChE activity following administration of an anticholinesterase agent may not be entirely responsible for the altered behavior. Reductions of 50% in brain AChE activity by eserine (1.28 mg/kg), compound 10854 (2.0 mg/kg), and carbaryl (10 mg/kg), with no overt signs of poisoning, caused rats to receive different percentages of administered shocks (60%, 43%, and 67%, respectively). In a gerbil study with phosdrin, behavioral decrements occurred only at dose levels that caused overt signs of poisoning (Mertens et al. 1974). The interest in behavior studies in the 1960s and 1970s prompted USEPA to convene a workshop for the purpose of assessing the capabilities of behavioral toxicology for predicting the neural and behavioral toxicity of chemicals (Geller et al. 1979). The challenge to the workshop participants included a need to (1) maximize the amount of information obtained from behavioral tests (i.e., delayed effects and recovery), (2) refine estimates of the dose of toxicant to the nervous system (i.e., neurochemical species, metabolism, and pharmacodynamic relationship), (3) determine whether a dose–response relationship exists and limitations of specific behaviors for the measurement of the dose response, (4) evaluate and allow for differences between individual test animals in susceptibility to effects of a toxicant, (5) consider whether enzyme induction or other effects or prior exposure to the test agent or other compounds affect the outcome of the behavioral test, (6) determine the importance of testing for potentiation (or diminishment) of effects from a given agent or the other compounds, (7) and determine the sensitivity of the behavioral test and the relevance to the human situation (Geller and Stebbins 1979).

The workshop was beneficial, but USEPA did not receive active support for these studies from other scientists. According to Karczmar (1984), a considerable number of studies were conducted after the workshop with war gases (i.e., sarin, soman, etc.) and physostigmine. Wolthius and Vanwersch (1984) reported the results of a behavioral study with teteaethylpyrophosphate (TEPP). In 1992, USEPA published Toxic Substance Control Act (TSCA) and Federal Insecticide Fungicide, Rodenticide Act (FIFRA) guidelines for conducting neurotoxicity studies that included studies involving behavior, motor activity, and neuropathology (USEPA 1991). Padilla et al. (1996) reported on the results of a study designed to determine whether blood AChE inhibition could be used to predict AChE inhibition in target tissue or behavioral effects. According to Padilla et al. (1996), whole-blood AChE inhibition by chlorpyrifos was a good predictor of brain AChE inhibition and changes in motor activity (% of control). Whole-blood AChE inhibition by aldicarb, carbaryl, and paraoxon were also found to be good predictors of changes in motor activity. The blood ChE–motor activity response curves varied considerably for the four pesticides; 60–100 mg/kg of chlorpyrifos (85% inhibition of whole blood AChE) were required to produce a 15% decrease in motor activity. For paraoxon and aldicarb, a 50% decrease in blood AChE was required to produce a noticeable change in motor activity; in the case of carbaryl, only 15%–20% whole-blood AChE inhibition was required to depress motor activity.

The studies by Padilla et al. (1996) did not consider the relationship between topically applied OPs, blood AChE inhibition, and motor activity. Knaak et al. (1989) developed dermal–ChE response data for several OPs and carbamate insecticides in the rat for the purpose of establishing reentry intervals, but did not examine the effect of AChE inhibition on motor activity.

VIII. Measured Variation of Esterase Activity in Human Plasma and Red Blood Cells

Callaway et al. (1951) studied the variation of plasma BChE and RBC AChE in 247 healthy adults. These were divided into three groups: males 18–30, fit for combat; males 18–69, no illness for 2 yr; females 25-74, civilians. There were no seasonal, age, or sex differences in enzyme activities. The percent variation of plasma BChE for the group was 21.7%. For RBC AChE, there was a 10.8% variation for service personnel and 15.4% for civilians as a group. Blood samples were taken from 10 male soldiers, aged 18–30, twice a week over 4 wk. For this fairly homogeneous group, the coefficients of variation were 8.5% (plasma, range 4.6%–11.8%) and 6.6% (RBC, range 3.9%–8.5%).

Augustinsson (1955) studied 201 healthy individuals, 141 males and 60 females, between 20 and 50 yr old. There was no sex difference in RBC AChE. One male was studied weekly for 2 yr. His RBC AChE was constant from week to week, and his plasma BChE varied about 6% from week to week. The coefficient of variation for plasma BChE of the male group was 14.7% and for females 22.9%. For RBC AChE, the male group coefficient of variation was 14.3% and for females 13.7%. There were 51 male and 31 female (plasma) and 49 male and 29 female (RBC) subjects. Augustinsson (1955) also compiled the plasma BChE and RBC AChE values from the literature: 1935–1955 for plasma and 1948–1955 for RBC AChE. For RBC AChE, the coefficient of variation was 18.6% for 1055 subjects of mixed sex and age. For plasma BChE, the variation was 21.9% for 3286 mixed individuals.

Gage (1955) measured the RBC AChE and plasma BChE of 19 normal individuals each month for 1 yr. The coefficient of variation was 12.8% (RBC) and 21.3% (plasma) for the group. Gage (1955) recommended that a 50% decrease in RBC AChE or plasma BChE compared to an individual preexposure baseline or the population average, whichever is higher, should require 'appropriate' action. Rider et al. (1957) studied plasma BChE and RBC AChE variation over 5 month in 400 males and 400 females who were healthy enough to give blood. There was no relationship between plasma and RBC enzyme activity. RBC AChE activity was not related to age; plasma BChE was higher with increasing age, and plasma BChE was higher in men than in women. RBC AChE activity was not statistically different between sexes, although males were consistently higher. RBC AChE showed about 11% interindividual variation for men and women and plasma varied about 20% (men) and 23% (women). Wetstone and LaMotta (1965) measured plasma BChE in 82 normal adults for up to 5 yr (373 total measurements). The intraindividual variation was 8.4%. Gage (1967), in a review of many studies, stated that if only one preexposure value was available, a 23% decrease in either RBC AChE or plasma BChE would be significant; and if a 'large' number of prior observations were available, a 16.5% decrease would be significant. This is similar to a table published by Callaway et al. (1951) in which, if one preexposure value was available, a 19.9% decrease in plasma BChE and a 15.3% decrease in RBC AChE would be significant. For four preexposure values, the differences are 15.7% (plasma) and 12.1% (RBC). A study of 1074 individuals from 25 families, including 15 pairs of monozygotic twins and 13 pairs of same-sex dizygotic twins all with the usual BChE phenotype, suggested that the level of plasma BChE activity appeared to be environmentally determined (Whittaker 1986).

Brock (1990) measured immunoreactive plasma BChE in six samples from 43 males and 51 females, aged 20–65 yr. The intraindividual variation in actual enzyme concentration was estimated to be 6.4%. The intraindividual variation in plasma BChE activity was 6.2%. Thus, in the absence of genetic conditions and disease, plasma BChE activity variation was caused by a 6.4% variation in actual enzyme level. These data suggested that intrinsic plasma BChE levels are fairly constant and that plasma ChE could be used as an OP exposure monitoring tool. Brock (1991) confirmed that interindividual variation in plasma BChE resulted from enzyme concentration and not inhibition of catalytic activity. Brock and Brock (1993) studied the interindividual variation in BChE in 650 males and 437 females using the Ellman method for enzyme activity (Ellman et al. 1961). These individuals were genetically defined as $ChE^{u}ChE^{u}$, $ChE^{u}ChE^{s}$ (phenotype U), or $ChE^{u}ChE^{a}$ (phenotype UA) (U = usual, A = abnormal, S = silent) by inhibition with dibucaine and fluoride. Interindividual variation ranged from 50% to 150% of the group mean with variation in the analytical method of <2%. Only about 0.1% of this variation was accounted for by sex and only 0.5% by age. Less than 10% of the variation was intraindividual variation. Overall, only 30% of the inter-individual variation was explained by ChE-1 phenotype, sex, body weight, and height. The authors concluded that "it may be difficult to compare ChE activities in various population groups" (Brock and Brock 1993). Further, Brock and Brock (1993) preposed a 'standardized' ChE activity calculated from ChE-1 phenotype, sex, and body-mass index when comparing population groups.

IX. Laboratory Measurement of Cholinesterases

Wilson et al. (1998) reviewed laboratory problems associated with determining plasma and RBC ChE activity in blood from workers exposed to OPs. The meth-ods measured activity after a specified incubation time (i.e., manometric method, ΔpH [Michel method], colorimetric, radiometric, and ELISA), except for the pH stat method which titrated released acetic or butyric acid during the course of the reaction. The pH stat method was capable of following the liberation of acid for a period of 1–5 min or longer. The method was used in the 1960s by Mengle and Serat (1975, personal communication) in California to obtain blood ChE data from workers exposed to OPs. The pH stat method was replaced by the Δ pH (Michel method) and colorimetric method in the 1970s because these methods were simpler to use and capable of analyzing more blood samples per hour. The basis for colorimetric methods (there are several manufacturers) for a clinical laboratory is the work by Ellman et al. (1961). Unfortunately, clinical laboratories have not standardized the esterase assays in use and do not use appropriate quality assurance techniques (Wilson et al. 1997). In addition, the same kit is used for RBC AChE and BChE. This use provides inaccurate measurement of one or both enzymes (Wilson et al. 1997). The pH stat method was never extensively used to develop kinetic data or determine ChE inhibition by toxicological laboratories engaged in the development of pesticides. Consequently, a limited amount of kinetic data is currently available on OPs used in agriculture. If the data had been developed, the information could be used in quantitative structure–activity relationship (QSAR) models for predicting inhibition and toxicity, physiologically based pharmacokinetic and pharmacodynamic models for risk assessment, and ligand docking models for developing and testing new pesticides. In only one case have ChE methods for monitoring workers been compared (Wilson et al. 1998).

In *Clinical Guide to Laboratory Tests* (Tietz et al. 1990, p. 126), anabolic steroids, carbamates, cyclophosphamide, estrogens, glucocorticoids, lithium, pancuronium, succinylcholine, oral contraceptives, OP insecticides, phenelzine, phenothiazines, iopanaic acid, ranitidine, synthetic estrogens, and testosterone are listed as interferences for plasma BChE measurement. For RBC AChE, only OP insecticides are listed (Tietz et al. 1990, p. 12).

There are two other methods for possible clinical use in determination of esterase inhibition by OP inhibitors. Polhuijs et al. (1997) published a method for reactivation of inhibited enzyme by fluoride ions with the production of a phosphofluoridate that is identified and quantified. This technique requires sophisticated technology, which probably puts it beyond a clinical laboratory. However, this technique should be very useful for special circumstances, e.g., confirming terrorist attacks with nerve agents.

Hansen and Wilson (1998) published a reactivation method using pyridine 2-aldoxime methochloride (2-PAM). Oxime reactivation of OP-inhibited esterases is known technology (Moore et al. 1995). The inhibited enzyme is measured, then incubated with 10 mM 2-PAM and remeasured. The increase in enzyme activity is the amount of OP inhibition. This is fairly complicated technology for a clinical laboratory and must be applied within 24 hr of exposure for accuracy. There are field kits for monitoring the blood esterases of exposed workers. These kits may be useful in heavy exposure situations, but appear to be not useful for "low" exposure situations (Atkins et al. 1998).

X. Selected Studies in Humans and Small Animals

A. Humans

Comroe et al. (1946a) administered DFP both i.m. and orally to seven myasthenia gravis patients. Plasma BChE was reduced to 0 in three patients, 1%–2% in three patients, and to 5% in one patient. RBC AChE was reduced to 5%–74% of pre-dose values, in six of seven patients to 5%–26%. These patients became nauseous; six vomited.

Grob et al. (1947a) administered DFP to more than 200 human subjects. The purpose of this study was to compare myasthenic subjects with normal subjects. DFP was administered i.m. in peanut oil (0.5–3.0 mg) and i.v. (0.5–2.0 mg) in water. With either route of administration and a single dose, plasma BChE (ACh was used as the substrate) was depressed 65%–95% and RBC AChE 5%–35% of the individual preexposure value. Plasma BChE maximum depression occurred within 1 hr, RBC AChE in 24 hr. A daily i.m. dose of 0.5–2.3 mg DFP caused an immediate and sustained depression of plasma BChE of 80%–95% and a progressive decline of RBC AChE of 5%–10%/mg of DFP/d. The maximum depressions for RBC AChE in three normal subjects were 1.3 mg/d (0.023 mg/kg) for 7 d, 76%; 1.4 mg/d (0.026 mg/kg) for 7 d, 75%; and 1.3 mg/d (0.022 mg/kg) for 8 d, 77%. These same subjects had plasma BChE depressions of 96%, 99%, and 96%, respectively. For the 25 subjects and all doses, plasma depression averaged about 89% and RBC depression averaged 50% for an average dose of 1.38 mg (0.021 mg/kg) for 3.8 d. Myasthenic subjects averaged about 91% plasma and 82% RBC depressions with an average dose of 0.80 mg (0.014 mg/kg) for 32.6 d.

With daily doses of 1.5–3.0 mg of DFP to humans, cholinergic symptoms usually appeared with 30% depression of RBC AChE. With a longer dosing period and 0.5–1.0 mg DFP/d, symptoms were absent or appeared and vanished with RBC AChE depressions of 75%. Regardless of percent depression, symptoms disappeared within 48 hr even in the absence of a rise in RBC AChE activity. After the last dose of DFP, plasma BChE recovered at an average of 14% (24 hr), 9% (24–48 hr), 6% (d 5), 3% (d 8), and 2% (d 15). RBC AChE recovered at 1%/d (Grob et al. 1947a).

In other studies, i.m. doses of DFP (1–3 mg) increased gut motility (Grob et al. 1947b) and caused electroencephalographic changes (Grob et al. 1947c). Changes in brain activity were not correlated with plasma BChE activity; but dosing over 2–3 d RBC AChE activity depression of 30%–40% was correlated with these changes. Doses administered beyond 3 d did not yield a correlation between RBC AChE and brain activities.

Brauer (1948) showed that plasma BChE is inhibited by TEPP *in vitro* to a greater extent than RBC AChE. Grob and Harvey (1949) studied the *in vitro* sensitivity of human plasma BChE, brain RBC AChE, muscle RBC AChE, and RBC AChE to parathion, DFP, neostigmine, and TEPP. Plasma BChE was ten fold more sensitive to parathion ($I_{50} = 1.5 \times 10^{-6}$ M) than RBC AChE ($I_{50} = 1.2 \times 10^{-5}$ M), to TEPP (5×10^{-9} M vs. 3.5×10^{-8} M), and to neostigmine (4×10^{-8} M vs. 8×10^{-7} M), but was 200 fold more sensitive to DFP (9.5×10^{-9} M vs. 4×10^{-7} M). Grob (1956) described three sarin (nerve agent) poisoning cases requiring atropine antidote. In these severe cases, plasma BChE and RBC AChE were depressed 70% and 85% or more, respectively. RBC AChE returned to normal in 3 mon, plasma BChE in 3–8 wk. For percutaneous absorption of a single dose of sarin, symptoms appeared when plasma BChE was depressed 85% and RBC

AChE 90% (Grob and Harvey 1958). Grob et al. (1950) reviewed the effects of parathion in man from poisoning cases. Four patients who recovered and two who died had 95% plasma BChE depression and about 86% RBC AChE depression. Subjects with moderate symptoms were depressed 90% (plasma) and 78% (RBC). As with DFP and TEPP, plasma was more rapidly depressed (more sensitive) than RBC AChE. Plasma BChE was not related to symptoms, whereas RBC AChE was related to symptoms up to about 3 d of exposure. Thereafter, RBC AChE remained depressed while symptoms disappeared, so tissue RBC AChE apparently recovers more quickly than RBC production. Grob (1950) showed that plasma BChE was about 10 times as sensitive as RBC AChE to parathion *in vitro*.

Moeller and Rider (1962) dosed volunteers (prisoners) orally with malathion, O-ethyl-O-*p*-nitrophenyl phenylphosphorothioate (EPN), and malathion plus EPN. Plasma and RBC enzymes were measured twice weekly for 78 d after dosing was complete. There were no clinical effects of 3 mg/d EPN, 8 mg/d malathion, 3 mg EPN plus 8 mg malathion/d, single doses of 6 mg EPN or 16 mg malathion, and single doses of 8 mg malathion plus 6 mg EPN, 3 mg EPN plus 16 mg malathion, nor was either blood enzyme depressed; 6 mg EPN plus 16 mg malathion/d for 42 d depressed plasma BChE and RBC AChE 15%–20% without symptoms. Finally, one group of five subjects was given 9 mg EPN/d for 56 d and another group of five subjects was given 24 mg malathion/d for 56 d. Enzyme depression was not seen until 2 wk after the first dose and maximum depressions of 30% (EPN group) and 25% (malathion group) occurred about 3 wk after the last dose. Plasma BChE was depressed sooner than RBC AChE, but to about the same extent. No clinical effects were noted.

Hayes et al. (1964) exposed volunteers to parathion and measured *p*-nitrophenol in urine and plasma BChE and RBC AChE. With 2% dust on the forearm for 2 hr over 5 d, no enzyme depression occurred. Exposure to 2% and 47.5% emulsions gave the same result. Using 7 lb of 2% parathion in a rubber suit, RBC AChE was depressed 16% and plasma BChE 56%. Exposure was for 7.5 hr and 7.0 hr on two occasions. Maximal excretion of p-nitrophenol was 507.6 µg. Exposure to parathion vapor caused no enzyme depression with an average excretion rate of *p*-nitrophenol of 92.1 µg/hr over 24 hr. Filter pads saturated with parathion and placed against the skin produced no RBC AChE depression and 20% plasma BChE depression with 2,370.5 µg of *p*-nitrophenol excreted over 24 hr.

Hartwell et al. (1964) exposed volunteers to parathion fogs. After the fourth exposure, depressions of 30% of both plasma BChE and RBC AChE were observed, and after the fifth exposure RBC AChE was depressed 98% and plasma BChE 83%. Research assistants inadvertently exposed were depressed 32% (RBC) and 57% (plasma) (subject A) and 52% (RBC) and 12% (plasma) (subject C) even though no *p*-nitrophenol was detected in urine. Hartwell et al. (1964) estimated that respiratory exposure was conservatively 10 fold as toxic as the dermal route.

Rider et al. (1969) studied the effect of parathion, systox, octamethyl pyrophosphoramide, and methyl parathion in human volunteers. Rider et al. (1969) defined minimal toxicity as 20%–25% RBC AChE or plasma BChE depression.

Volunteers were dosed orally over 30 d. The doses per day to reach 20%–25% plasma or RBC enzyme depression were parathion, 7.5 mg; systox, 6.75–7.125 mg; OMPA, 1.5 mg; and methyl parathion, >19.0 mg. Plasma BChE was depressed more quickly, except for OMPA. The 1.5 mg/d dose of OMPA produced a 23.5% plasma BChE after 30 d. RBC AChE was more sensitive to OMPA (34% depression 5 d after the last dose).

Ludwig et al. (1970) exposed 5 volunteers to fogged chlorpyrifos at doses used for mosquito control. Neither RBC AChE or plasma BChE depression occurred. Sherman et al. (1988) administered 60 μg/kg of neostigmine bromide (a carbamate) orally to Alzheimer patients. Maximum depression of both enzymes occurred in 60 min, but plasma BChE was twice as inhibited compared to RBC AChE and plasma was more variable (4%–21% in 60–150 min) compared to RBC AChE.

Table 11 presents a summary of human esterase depression and the appearance of symptoms.

B. Small Animals

Hawkins and Mendel (1949) studied the *in vivo* inhibition of rat plasma BChE and RBC AChE by oral doses of Nu-1250, a prostigmine analog. RBC AChE from plasma was inhibited in symptomatic rats, whereas plasma BChE was uninhibited. In the same study, human RBC AChE was 100% inhibited by 10^{-5} M Nu-1250 *in vitro* whereas plasma BChE was 29% inhibited. Freedman et al. (1949) injected adult male rats s.c. with 1–2 mg/kg DFP in saline. With 2.0 mg/ kg of DFP, plasma BChE, RBC AChE, and brain AChE were almost totally inhibited and 25% of the rats died in the first hour. In the survivors, plasma BChE regenerated 40% in 24 hr and reached control levels in 7 d. RBC AChE remained inhibited for 48 hr and then regenerated at about the same rate as RBCs. Brain AChE recovered over 10 d to about 45% of normal with no lag. The most consistent correlation between symptoms and enzyme inhibition was for brain AChE, followed by RBC AChE. Plasma BChE failed to correlate with symptoms (Freedman et al. 1949).

Frawley et al. (1952) studied the toxicity and ChE inhibition of DFP, parathion, TEPP, EPN, OMPA, and E-838 (diethoxythiophosphoric acid ester of 7-hydroxy-4-methyl coumarin) in rats, mice and guinea pigs. Subacute doses were administered to rats in lab chow. Acute doses were administered to 24-hr-fasted rats, mice, and guinea pigs by stomach tube. Acute oral LD_{50}s with the male rat were TEPP, 2.0 mg/kg; DFP, 13.5 mg/kg; OMPA, 13.5 mg/kg; parathion, 30.0 mg/kg; E-838, 42.0 mg/kg; and EPN, 91.0 mg/kg. With the male rat, subchronic feeding studies yielded more RBC AChE inhibition compared to plasma BChE with all compounds. At 25 ppm for 2 wk, only DFP inhibited brain ChE (about 55%). Feeding 5 ppm of each compound showed no brain AChE inhibition; OMPA inhibited plasma BChE 32% at 2 wk and 10% at 8 wk. No other plasma inhibition occurred at 5 ppm. RBC AChE showed inhibition with 5 ppm DFP (17%), parathion (54%), and OMPA (47% at 2 wk and 64% at 8 wk). Fraw-

Table 11. Human blood esterase depression and symptoms.

Reference	Compound	% Depression RBC	% Depression Plasma	Symptoms
Comroe et al. (1946a,b)	DFP	75	100	Nausea
Grob et al. (1947a)	DFP	35	95	Nausea
Grob et al. (1950)	Parathion	86	95	Death
		78	90	Moderate symptoms
Grob and Harvey (1958)	Sarin	90	85	Symptoms appear
Moeller and Rider (1962)	Malathion	25	25	No symptoms
	EPN	30	30	No symptoms
Hayes et al. (1964)	Parathion	16	56	No symptoms
Hartwell et al. (1964)	Parathion	98	83	Symptoms
Rider et al. (1969)	Parathion, systox, octamethyl pyrophosphora mide, methyl parathion	25	25	No symptoms
Ludwig et al. (1970)	Chlorpyrifos	0	0	No symptoms

ley et al. (1952) unequivocally concluded that if humans reacted to OPs similarly to rats, RBC AChE would be the best procedure for detecting poisoning or exposure to anticholinesterases.

Reiter et al. (1975) studied inhibition and recovery of blood ChEs in bonnet and rhesus monkeys. Parathion was administered orally in single doses concealed in fruit. The peak inhibition with all doses, sexes, and species occurred in about 6 hr. At 2.0 mg/kg, plasma BChE was about 80% inhibited and RBC AChE 60% inhibited. Both enzymes recovered almost to normal over 2 wk at about the same rate. Visual discrimination performance deterioration coincided with peak enzyme inhibition (Reiter et al. 1975).

Pasquet et al. (1976) studied the blood enzyme inhibition and toxicity of phosalone, azinphosmethyl, and parathion in rats. The oral LD_{50}s were 10 mg/kg (parathion), 25 mg/kg (azinphosmethyl), and 120 mg/kg (phosalone). Dermal toxicities to the female rat were approximately 1530 mg/kg (phosalone), 380 mg/kg (phosalone oxon), 90 mg/kg (azinphosmethyl), and about 8 mg/kg (parathion). After oral dosing, all three compounds were clearly better inhibitors of plasma BChE compared to RBC AChE (Pasquet et al. 1976). Reiter et al. (1975) studied inhibition and recovery of blood ChEs in bonnet and rhesus monkeys. Parathion was administered orally in single doses concealed in fruit. The peak inhibition with all doses, sexes, and species occurred in about 6 hr. At 2.0 mg/kg, plasma

BChE was about 80% inhibited and RBC AChE 60% inhibited. Both enzymes recovered almost to normal over 2 wk at about the same rate. Visual discrimination performance deterioration coincided with peak enzyme inhibition (Reiter et al. 1975).

XI. Field Studies

Sumerford et al. (1953) studied 258 persons exposed to OPs in and around Wenatchee, WA, in apples. Seventeen of these were orchard 'workers', i.e., not applicators. Their mean plasma BChE was 0.87 ΔpH/hr preexposure; 0.81 ΔpH/hr exposure; 0.89 ΔpH/hr recovery; and 1.17 ΔpH/hr postrecovery (only one person was measured postrecovery). RBC AChE was depressed from 0.76 to 0.67 and recovered to 0.71 ΔpH/hr. Mixing plant personnel plasma BChE was reduced from 0.96 to 0.39 and recovered to 0.78, and their RBC AChE was reduced from 0.82 to 0.46 and recovered to 0.61. Increased complaints of illness (headache, nausea, fatigue) were generally linked to exposure and lowered RBC AChE and plasma BChE values. Bogusz (1968) studied 43 subjects applying metasystox (S-[2-ethylthioethyl] O,O-dimethyl phosphorothioate), ekatin (thiometon; S-{2-(ethylthio) ethyl} O,O-dimethyl phosphorodithioate), folithion (fenitrothion; O,O-dimethyl 0-4-nitro-M-tolyl phosphorothioate), lebaycid (fenthion; O,O-dimethyl 0-[3-methyl-4-(methylthio) phenyl] phosphorothioate), and dipterex (trichlorfon; dimethyl (2,2,2-trichloro-1-hydroxylethyl) phosphonate). At the end of 24 d of exposure, RBC AChE was depressed 25% and plasma BChE 68%, on average.

Knaak et al. (1978b) measured the RBC AChE and plasma BChE in field workers and nonfield workers in California. Histograms (frequency vs. activity) of RBC and BChE ChE activity were constructed for field workers. Male workers had RBC AChE activity (mean value) equivalent to Sacramento blood bank donors and field workers that had not worked in the field during the past 30 d. Males living in agricultural areas but not working in the fields had slightly higher RBC AChE activity. Field workers had plasma BChE activity values equivalent to Sacramento blood bank donors and workers that had not worked in the field during the past 30 d. Males living in agricultural areas but not working in the field had slightly lower plasma BChE activity values. Activity values (μM of -SH min^{-1} mL^{-1}) for male field workers ranged from 14.3 to 40.7 for AChE and 3.2 to 18.4 for BChE. Workers with low AChE and BChE values may be at greater risk than those with higher values. Worker exposure to OPs resulting in illnesses causes a shift in the mean (AChE values) to lower values and a corresponding shift in the minimum and maximum values closer to the mean value.

Knaak et al. (1978c) investigated an illness episode involving 118 workers exposed to dialifor (Torak®) and phosalone reentering grape vineyards. Workers were exposed to dislodgeable foliar residues of dialifor of 0.54 μg/cm^2 on Aug. 30 and to 0.14–0.16 μg/cm^2 on Sept. 7–9, just before reporting ill. RBC AChE and plasma BChE were depressed to about 45% (RBC) and 50% (plasma) of control with dislodgeable residues estimated to have been 2.2 μg/cm^2. Phosalone residues of 0.7 μg/cm^2 and phosalone oxon of 2.2 ppm total did not depress RBC

AChE of workers (Knaak et al. 1978c). Knaak et al. (1978a) also studied workers reentering a citrus grove treated with phosalone 21 d earlier and another grove treated with phosalone 14 d earlier. Phosalone residues were 3.6 µg/cm^2 after 21 d and declined to 2.9 ± 0.24 µg/cm^2 during the 3-d work period (phosalone oxon, about 0.50 µg/cm^2). For the 14-d reentry grove, phosalone residues were 2.6 ± 0.48 µg/cm^2 on the day of entry and 2.42 ± 0.09 µg/cm^2 on the last work-days (phosalone oxon, 0.19–0.27 µg/cm^2). RBC AChE was not different at the 5% level comparing field workers to controls. Plasma BChE was statistically lower in exposed versus controls in both studies. This reduction in plasma BChE was described as small, and a 21-d reentry period was recommended for phosa-lone on citrus (Knaak et al. 1978a).

Areekul et al. (1981) studied OP suicide cases and found that serum plasma BChE was depressed to zero in two fatal cases. Areekul et al. (1981) also mea-sured serum plasma BChE and RBC AChE in exposed and nonexposed factory workers. Serum plasma BChE was depressed about 30% in both groups, whereas RBC AChE was not depressed in either group. Areekul et al. (1981) concluded that serum plasma BChE could help confirm exposure but was not reliable to fol-low the clinical course of intoxication. Hassan et al. (1981) correlated serum plasma BChE recovery with clinical recovery of two patients. However, these authors also erroneously stated that serum plasma BChE is used to estimate true ChE hydrolytic activity at the neural junction. Duncan et al. (1986) reviewed exposure studies that used plasma BChE to monitor OP exposure. They sug-gested the use of plasma BChE and a calculation of number of workers below the group mean to estimate ill workers.

Drevenkar et al. (1991) studied mixers, applicators, field workers, mechan-ics, and a manager, a technologist, and a housekeeper employed on a farm where methidathion, vamidothion (O,O-dimethyl S-[z-[{1-methyl-2-(methylamino)-2-oxoethyl}thio]ethyl]ester) (2275-23-2), and azinphosmethyl were used. In 80 postexposure samples, plasma BChE was inhibited from 31% to 48% and for 68 samples 0% to 30%. Of 95 urine samples, 17 contained no metabolites. Four sub-jects with no metabolites in the urine had depressions of plasma BChE of 31%–48%. Drevenkar et al. (1991) concluded that both enzyme and urinary metabo-lites should be monitored for absorption of OPs.

McCurdy et al. (1994) studied urinary alkylphosphate levels, plasma BChE, and RBC AChE in workers reentering a peach orchard treated with azinphosme-thyl 30 d previously. Dislodgeable foliar residues were measured. Baseline enzyme values for each subject were not established. RBC AChE had decreased 7% by d 3 of exposure and declined 19% over 6 wk. Plasma BChE declined 9% by d 3 and 12% over 6 wk. The authors concluded that urinary metabolites are better than enzyme measurements for confirming exposure and that RBC AChE measurements were better for detecting exposure to azinphosmethyl (McCurdy et al. 1994).

O'Malley and McCurdy (1990) investigated a reentry incident involving phosalone applied to grapes. Ten crew members were admitted to the hospital with complaints of headache, nausea, abdominal pain (cramps), and dizziness.

One worker, the first to be admitted, also complained of vomiting, lacrimation, salivation, and disorientation. These workers were exposed to 0.3–2.0 $\mu g/cm^2$ phosalone and 0–0.29 $\mu g/cm^2$ phosalone oxon. For the entire crew, median plasma ChE inhibition was 64.9% and for RBC AChE 52.5% compared to general population means. For the 20 crew members not admitted to the hospital, plasma BChE depression averaged 57.7% and for RBC AChE 51%. For the 10 workers admitted to the hospital, RBC AChE depression averaged 61.9% and plasma BChE 80.7%. Bradycardia, symptomatic of OP poisoning, was evident in some subjects. The assertion that the cholinergic symptoms observed resulted from phosalone was not supported by residue levels and etiology. Subject no. 1 displayed significant enzyme depression after less than 2 d of exposure to low residues. Crews were exposed to higher levels of OP residues in previous experiments with no effect on RBC AChE and slight effects on plasma BChE (Knaak et al. 1978a,c). 2-PAM was not given. According to Namba (1971) and Namba et al. (1971), had 2-PAM been given, the symptoms might have disappeared and RBC AChE increased due to reactivation. No urinary metabolite analyses were run for confirmation. Phosalone was the only anticholinesterase compound found in the field and, consequently, was tagged as the culprit (O'Malley and McCurdy 1990). The ChE depressions were large and the dose of phosalone and its oxons, both relatively nontoxic, were very low.

Hayes (1969) warned, "No discussion of mortality, morbidity, storage, or any other effect of chemicals is meaningful except in terms of dosage." For this reason, doubt is associated at once with any pesticide health case alleged to result from exposure that is trivial in comparison with what people ordinarily withstand without inconvenience. The real harm of attributing diseases of truly unknown cause to pesticides without adequate evidence is that the search for the true cause may be abandoned. It would have been a pity, for example, if the study of the viral origin of poliomyelitis had been abandoned because one person thought the disease was caused by DDT.

XII. Other Human Conditions and Blood Enzyme Depression

A. Plasma Cholinesterase in Disease

Plasma BChE is depressed in liver disease, anemia, acute infectious diseases, cardiac failure, and renal disease (Grob et al. 1947a). Long (1975) listed cirrhosis, chronic toxic or viral hepatitis, malignancies, hepatic or obstructive jaundice, decompensated heart disease, allergic disease, and pregnancy associated with plasma BChE depression. Plasma BChE is decreased (LaMotta et al. 1957) and the 'isozyme' pattern altered in liver cirrhosis and metastatic liver cancer (Matsuzaki et al. 1980a,b; Stefanini 1985). Gene amplification of both AChE and BChE genes occurs in blood disorders (Lapidot-Lifson et al. 1989). Plasma BChE isozyme patterns have been used to differentiate liver cirrhosis from chronic hepatitis (Hada et al. 1990). Immunological determination of amniotic RBC AChE and plasma BChE can be used to distinguish pregnancies complicated by anencephaly, open spina bifida, encephalocele, and miscarriage from ventral wall defects (Loft 1990; Loft et al. 1991). Plasma BChE in cerebrospinal fluid may also be useful for diagnosing Guillain–Barré syndrome and meningitis (Atack et al. 1987). Whittaker (1986) lists 38 reasons for decreased plasma BChE from pregnancy to x-ray therapy to tetanus. A reduction to 25% activity in a disease condition usually heralds death (Whittaker 1986).

We suggest the article by Layer and Willbold (1995) for a review of ChE function in animals and humans. Layer and Willbold review ChEs in various organs, in developing embryos, and possible involvement in human disease including birth defects, Alzheimer's disease, and cancer.

B. Genetic Variation

La Du et al. (1991) have suggested nomenclature for the nine genetic variants of human serum plasma BChE. These variants and their mutations are apparently encoded by one gene (La Du et al. 1991; Lockridge 1990). For the purposes here, the genetic variants are rare and become part of a subject's preexposure baseline during workplace monitoring. Some genetic variants are resistant to inhibition by dibucaine and to positively charged inhibitors (Prestor and Simeon 1991) while others have lower activity than normal (see Table 3).

The human BChE gene is present in a single copy (Arpagaus et al. 1991). Genetic variants are mutant forms caused by point mutation (atypical, fluoride-resistant), frame-shift mutation (silent), and polymorphic site (K-variant) (La Du et al. 1990). The characteristics of the usual plasma BChE (96 of 100 persons) have been reviewed by Lockridge (1990). This esterase hydrolyzes succinylcholine, procaine, 2-chloroprocaine, meprylcaine, isobucaine, tetracaine, aspirin, heroin, cocaine, methylprednisolone acetate, aprophen, mivacurium, and bombuterol (Lockridge 1990). It is possible that ethylcocaine may be produced by this and similar esterases in individuals consuming cocaine and ethyl alcohol (Dean et al. 1991). The highest turnover number (mol hydrolyzed/min/active site) for these drugs is for aspirin (7200) and the lowest are for aprophen (0.12) and cocaine (0.4) at saturating substrate concentration. BChE may be significant in cocaine detoxification (Lynch et al. 1997) and has been suggested for therapeutic use in cocaine intoxication (Mattes et al. 1997).

There is no link to any disease from the genetic variants. Persons with silent plasma BChE are healthy and appear to have a normal lifespan (Hodgkin et al. 1965). No physiological function for plasma BChE is known, although high levels in the embryo and in the newborn suggest a role in development (Ecobichon and Stephens 1972; Zakut et al. 1991), and a role in lipoprotein metabolism is suggested by a high level in obesity and a low level in malnutrition (Lockridge 1990).

C. Possible Dietary Effects on Blood Esterase Activity

In rats, plasma BChE activity is decreased by dietary fats. This decrease ranges from 9% to 45% dependent on the fat fed (Van Lith et al. 1990). Fish oil may also reduce plasma BChE in humans (Houwelingen et al. 1987).

While trying to measure the pesticide residue level of OP compounds with a ChE assay, Orgell et al. (1958) discovered the potato controls were inhibitory. A similar experiment by Heath and Park (1953) disclosed a ChE inhibitor in white clover. Galanthamine, an alkaloid ChE inhibitor, is found in *Galanthues* spp. (Domino 1988). For the potato, the inhibitor(s) was concentrated in the peel (Orgell et al. 1958). Harris and Whitaker (1959) used potato peel extracts to inhibit plasma BChE from atypical-homozygote, normal, homozygote, and heterozygote human genotypes. Subjects had been classified genetically by dibucaine number (inhibition) and, using a potato peel extract, normals were inhibited 77%–83%, heterozygotes 53%–67%, and atypical 16%–24%. Heterozygotes are about 4% of the population, normals about 94%, and atypical about 1%. Orgell et al. (1959) studied weedy and cultivated species for anti-plasma BChE activity. Strong inhibitory activity was found in the edible portion of potato and eggplant and in tobacco leaves. Orgell and Hibbs (1963) extended their work to classify plasma BChE inhibition to potato varieties and ascribed the inhibition to leptine I and II and solanine. Orgell (1963) also tested alkaloids, glycosides and other natural substances against plasma BChE. Of 139 compounds, 88 produced less than 10% inhibition; these included ubiquitous compounds (for humans) such as caffeine, quercetin, and nicotine. Eleven compounds produced 10%–31% inhibition; none were very common. Thirty-four were very inhibitory; these included leptine, solanine, demissidine, bishydroxycoumarin, and naringenin. Phytostigmine was most inhibitory with an I_{50} of 7×10^{-9} M. Leptine and solanine had I_{50} values of 4×10^{-6} M and 5×10^{-6} M, respectively.

Sinden and Webb (1972) found levels of 14–650 mg/kg alkaloid in various potato varieties. Levels above 260 ppm have caused human illness (Wilson 1959). Patil et al. (1972) evaluated the toxicity of solanine in mice and rabbits. The i.p. LD_{50} was 32.3 mg/kg for mice. Rabbits given 20 mg/kg i.p. usually died in 24 hr. Rabbits given 30 mg/kg i.p. died in 50 min. In the female rabbit, plasma BChE was 45% and RBC AChE 34% inhibited after 24 hr. Atropine is also an inhibitor of plasma BChE (Vincent and Parant 1956). Atropine, morphine, procaine, amphetamine, and methylene blue appear to be competitive inhibitors of plasma BChE (Goldstein 1951).

Potato glycoalkaloids are not nontoxic natural compounds. Based on human death and poisoning cases, toxic doses range from 2 to 5 mg/kg and lethal doses from 3 to 6 mg/kg. By comparison, other oral LD_{50}s for known poisons are strychnine (5 mg/kg), arsenic (8 mg/kg), parathion (10 mg/kg), azinphosmethyl (25 mg/kg), and phosalone (120 mg/kg) (Morris and Lee 1984; Pasquet et al. 1976). These alkaloids cross the gut wall, and the maximal uptake appears to occur in about 12 hr (Morris and Lee 1984). The dose in a 500-g serving of potatoes is from 20 to 120 µg/g or about 0.2–0.9 mg/kg (Morris and Lee 1984). Peak plasma levels of α-solanine and α-chaconine were 5.1 and 6.0 hr, respectively, after a potato meal, with plasma half-lives of 11 and 19 hr, respectively (Hellenäs et al. 1992).

Alozie et al. (1978) studied the toxicity and rat isozyme ChE inhibition by α-chaconine, another potato alkaloid. Brain RBC AChE activity was dose related, 10, 30, and 60 mg/kg yielding 18%, 55% and 79% inhibition, respectively, compared to controls. Heart RBC AChE and plasma BChE were inhibited 61% and 51%, respectively, with 10 mg/kg and also with 30 mg/kg. Alozie et al. (1978) ran PAGE (polyacrylamide gel electrophoresis) on tissues of rats dosed i.p. with 10, 30, 60, and 90 mg/kg of α-chaconine. Band 1 of brain was eliminated by 30 and 60 mg/kg of α-chaconine and band 2 was reduced; band 3 was not affected. With acetylthiocholine (AcThCh) as a substrate, 5–6 bands were shown in control plasma. Activity band 3 was absent in animals treated with α-chaconine. With α-naphthylacetate as substrate, two additional bands were evident in rat brain. In rats given 60 mg/kg, bands 7 and 5 were eliminated; band 2 was not inhibited, bands 1, 3, and 4 were reduced from moderate to weak, and band 6 was reduced from strong to moderate. These are *in vivo* results.

In vitro incubations of electrophoretic gels with 10^{-4} M α-chaconine and acetylthiocholine as substrate showed that brain activity bands were all inhibited to some extent. For plasma and erythrocytes, not all bands were inhibited and the extent of inhibition for bands that were inhibited was different from brain (Alozie et al. 1978).

α-Chaconine inhibition of plasma ChE, based on rat *in vivo* results, appeared to be (1) potent and (2) irreversible (Alozie et al. 1978). The inhibition of plasma BChE and RBC AChE by tacrine (1, 2, 3, 4-tetrahydro-9-acridinamine) is prolonged *in vivo* and has been shown to be reversible *in vitro* (Dawson 1990). Dawson (1990) showed RBC AChE inhibition by tacrine is independent of time, is reversed by dilution, decreases with increasing substrate concentration, and is reversed by dialysis. An experiment with these criteria for a reversible inhibitor (and conversely for irreversible inhibition) has been conducted for potato alkaloids (Nigg et al. 1996). α-Chaconine and α-solanine were found to be reversible inhibitors of human plasma BChE (Nigg et al. 1996). Based on the discovery of potent AChE inhibitors (alkaloids, huperzine A, and huperzine B) from *Huperzia serrata*, a Chinese folk medicine, we are likely to discover other "natural" anticholinesterases in folk medicines and foods (Tang et al. 1988; Kozikowski et al. 1991). Huperzine A is three fold more potent than physostigmine against AChE but is less potent than physostigmine against BChE (Kozikowski et al. 1991).

Potato alkaloids appear to selectively inhibit plasma BChE compared to RBC AChE, as do curare alkaloids (Harris and Harris 1944). However, galanthamine, an alkaloid of *Galathus nivalis* (common snowdrop), is a selective inhibitor of RBC AChE compared to plasma BChE (Thomsen and Kewitz 1990). Atypical plasma BChE variant enzyme is differentially inhibited by solanidine and solanine (Harris and Whittaker 1962) and is not reactivated with 2-PAM like the usual enzyme (Neville et al. 1990). The inhibition of the atypical and typical plasma BChEs is the same with DFP and TEPP, but the atypical phenotype is more resistant to inhibition by neostigmine, physostigmine, and other non-OP inhibitors (Kalow and Davies 1958). Some physostigmine analogs differentially inhibit either RBC AChE or plasma BChE (Atack et al. 1989). Compared to RBC AChE, the usual plasma BChE is differentially inhibited by physovenines and its analogs *in vivo* in the rat (Yu et al. 1991). Plasma BChE averaged 8.1 fold the sensitivity to six physovenines as RBC AChE. For compounds not included in this average, one was equally inhibitory, one ((-)-phenylcarbamate) was 62.5 fold more inhibitory to RBC AChE and one ((-)-cumylcarbamate) was 233.9 fold more inhibitory to plasma BChE (Yu et al. 1991).

Reversible inhibitors like physostigmine (eserine) (Grob et al. 1947a) and carbaryl (Carter and Maddux 1974) result in an *in vitro* reduction in plasma BChE or RBC AChE. This effect was also obtained *in vivo* (Grob et al. 1947a; Harvey et al. 1947). The anticholinesterase action of solanine is apparently antidoted by atropine (Patil et al. 1972).

Although blood esterase inhibition by dietary factors may be rare (probably does not exist with the potato), it should always be a consideration in a testing program. Particularly important is differential inhibition of esterases by dietary compounds. For example, the consumption of one 'wrong' potato can produce OP poisoning symptoms, but probably little or no blood esterase inhibition would be seen.

XIII. Historical Esterase Monitoring Recommendations

RBC AChE and plasma ChE have been used to assess the physiological course of OP exposure (Gage 1967; Hayes 1983; Peakall 1992) (Table 12). Ganelin (1964) suggested that a 25% depression in either RBC AChE or plasma BChE requires investigation of safety precautions. Zavon (1965) recommended that a depression of plasma BChE without depression of RBC AChE was a warning of possible exposure. A 40% decrease in RBC AChE was a danger signal and a 60% depression was the suggested trigger point for removal from exposure to OP compounds (Zavon 1965). Gage (1967) recommended RBC AChE inhibition 30% below an established preexposure level as a criterion for removing a worker from further exposure. Gage (1967) calculated that, based on normal variation in both enzymatic activities and with solid preexposure means, a 17% reduction in activity was significant; that is, a 17% reduction in either plasma BChE or RBC AChE was actual depression. Gage (1967) suggested that both plasma BChE and RBC AChE could be reduced to 70% of a preexposure value without the slightest risk of toxic effects.

Namba (1971) stated that when more than 50% inhibition of either plasma BChE or RBC AChE occurs, signs and symptoms of OP poisoning will be present. Although Namba (1971) theoretically preferred the measurement of RBC AChE, he recommended plasma BChE because of its convenience and stability after sampling. Namba and Hiraki (1958) had previously recommended plasma BChE for diagnosing OP poisoning because RBC AChE is "complex" to measure and "because of a relatively long time required in becoming proficient in

Table 12. Historical recommendations for worker safety based on cholinesterases.

Reference	% Depression		Recommendation
	RBC AChE	BChE	
Ganelin (1964)	25	25	Investigate safety procedures
Zavon (1965)	40	—	Danger signal
	60	—	Remove worker
Gage (1967)	30	—	Remove worker
Namba (1971)	50	50	Symptoms present
Vandekar (1980)	25	—	Better safety precautions Lighter work schedule
	37.5	—	Remove worker
	50	—	
Morgan (1982)	25	25	Excessive exposure

the operation." Namba and Hiraki (1958) were treating ill workers with 2-PAM when they made this recommendation. Namba (1971) and Namba et al. (1971) classified OP poisoning into four categories: (1) latent poisoning—no clinical manifestations, plasma BChE activity depressed no more than 50%; (2) mild poisoning—patient can walk, fatigue, headache, dizziness, sweating, salivation, tightness in chest, abdominal cramps, diarrhea, plasma BChE activity depressed 50%–80%; (3) moderate poisoning—patient cannot walk, generalized weakness, difficulty in talking, other signs as in mild poisoning, plasma BChE 80%–90% depressed; and (4) severe poisoning—patient is unconscious, miosis, no pupillary reaction to light, muscular fasciculations, cramps, secretions from mouth and nose, cyanosis, plasma BChE more than 90% depressed. These recommendations were based on patients who were suicidal, had accidently ingested an OP pesticide (primarily parathion or methyl parathion), or were very heavily exposed occupationally. Long (1975) suggested that a depressed plasma BChE without clearcut exposure is reason to test for a genetic variant. Murphy (1980) stated that either plasma or RBC ChE was reasonably well correlated with severity of exposure and poisoning, but measurement of RBC AChE was considered more reflective of nervous tissue activity. Vandekar (1980) recommended that a 25% reduction in RBC AChE was reason for better adherence to precautionary measures, a lighter spraying schedule with 37.5% reduction and a withdrawal from exposure with a 50% reduction. A 12.5% reduction of RBC AChE in a group of workers was recommended as a signal that better precautionary measures should be instituted. RBC AChE was recommended over plasma BChE because RBC AChE is associated with a more specific biological response. Morgan (1982) suggested that a 25% depression of RBC AChE or plasma BChE compared to a preexposure baseline is strong evidence of excessive exposure.

There is no agreement as to the relationship between the inhibition of blood esterases and the health status of the individual (Pesticide & Toxic Chemical News 1991). Part of this disagreement is related to the technical facts we have presented. For example, is there differential OP inhibition of RBC AChE compared to plasma BChE compared to brain AChE with any specific OP compound? Based on the data, this possibility exists. Do individuals vary in their level of esterases? Unequivocally, they do. Is it possible that the clinical or experimental measurement of esterases is wrong? Yes.

XIV. Recommendations for a Blood Esterase Monitoring Program

Before an employee is allowed to apply pesticides, a complete medical history that includes a blood chemistry should be obtained. Employees with significant medical problems should not apply pesticides, and employees taking drugs that interfere with blood esterase tests should not apply esterase-inhibiting pesticides.

Measurement of a worker's blood esterases and subsequent comparison to a population range of esterase activity is a very poor practice for a routine monitoring program. Population ranges for blood esterase measurements give a broad set of activities for comparison and this set applies to an individual only marginally. Population ranges are used when an individual's baseline (unexposed value) is not known, and this usually occures with suicide and emergency room cases.

Because individuals differ from one another in their blood esterase levels, a meaningful assessment of exposure to esterase-inhibiting pesticides requires that an individual's normal esterase level must be determined. At least two and preferably three determinations of RBC AChE and BChE before pesticide exposure should be made. The difference between the values (highest and lowest) should be no more than 15% of the high value. Determinations should be made no less than 48 hr but not more than 15 d apart (Nigg and Olexa 1986).

It is ideal if workers were not exposed to OPs for a minimum of 30 d before establishing baseline values, but for many farming operations this may not be possible. Workers may be hired and their pesticide exposure history is not clear, or work schedules may not permit a 30-d nonexposure period. Assuming a weekend off, remove the worker from exposure on Monday, take one measurement on Wednesday and a second measurement on the following Monday. If the two values do not vary by more than 15% of the high value, average them for the person's baselines for BChE and RBC AChE.

Check out the testing laboratory. Have they run this test before? Are they willing to run one blood sample five times to show you how much their laboratory procedures vary? Do they run esterase determinations at one and always the same temperature? Which kit do they use? What is the population range for this kit? How quickly will reports be available? What is the report format? Does the laboratory provide a report that includes the population range for each enzyme? Will they include the person's baseline values on each report? Is the report easily read and interpreted? These questions avoid frustrations later when employees begin asking questions about their reports. Each employee is entitled to a copy of their own report. The local health department may agree to draw blood for a nominal fee. The health department can also get a favorable price for the tests and can provide physician interpretation and other health services.

There are two ways to monitor workers once the baselines are obtained. The first makes almost no decision on the extent, duration or level of exposure. Workers are monitored at some specified interval. For instance, a vegetable grower applies an OP pesticide 3 d/wk. A monitoring program should begin with the baseline followed by monitoring every Friday for 1 mon. If there is no esterase depression of 20%–30% monitor every second Friday for 1 mon. If there is still no esterase depression, set up a monitoring program for every third Friday. For other workers who are exposed to the same pesticides and in about the same amounts, also monitor every third Friday. For persons exposed 1 d/wk, monitor the day after the exposure. For persons who spray daily, monitor at least every second Friday. Do not monitor on Mondays because this means at least 1 d without exposure before esterase measurement, allowing recovery of enzyme activity. The numbers you receive from the clinical laboratory will give you guidance on how often to monitor.

Occasionally, there are accidents or workers complaining about symptoms (headache, upset stomach, fatigue) or suspicions that workers are not being careful. For accidents, workers should be monitored within 24 hr. For workers complaining about illness, monitoring should be performed on the same day and the worker should be removed from exposure. Monitor careless workers more often.

A blood profile should be run every 6 mon on each employee working with pesticides. This blood profile should be interpreted by a physician. This is not a test for pesticide exposure or illness caused by pesticides. It does, however, provide a general health picture and will detect gross pathological conditions. Again, persons with significant health problems should not apply pesticides.

Possible esterase scenarios are as follows.

1. Plasma BChE is depressed, but RBC AChE is not depressed.

This is a warning that exposure may be occurring. Investigate work and hygiene practices. Is the worker taking a prescribed drug that lowers plasma ChE? Does the worker have a drinking problem? Is there a related health problem? This person should be monitored more frequently, but there are so many substances and conditions that depress plasma ChE that removal from exposure is not warranted unless plasma ChE depression continues or RBC AChE becomes depressed. If, in a subsequent test, RBC AChE shows a 15%+ depression and plasma BChE continues downward, remove the worker from exposure. Here, the 20%–30% RBC AChE criteria does not apply. Because RBC AChE has become depressed, the situation is progressing and the person should be protected. In this case, a very thorough review of work practices is in order. Retest in 1 wk. If the values for RBC AChE and plasma BChE have returned to the person's baseline range, they may return to work.

2. Plasma BChE is not depressed. RBC AChE is depressed 20%–30%.

Remove the worker from exposure for 1 wk and retest. This worker may have been exposed in the near past. Because plasma BChE recovers more quickly than RBC AChE, RBC AChE can be depressed but BChE not depressed at test time. Atypical plasma BChE may be involved in this case, which means plasma BChE may not be as sensitive an indicator as with a normal genotype. Plasma BChE generally is considered to be 10 fold as sensitive to OP pesticide inhibitors. However, based on other inhibitors, RBC AChE could be more sensitive. OPs are not assessed for this possibility when registered.

Thoroughly counsel a worker in this situation to wear clean clothes, change into clean footwear, and to double-wash their hat or discard it and wear a new one. Leather boots and canvas/leather sneakers are not recommended for pesticide workers because they soak up pesticides like a sponge. Leather and canvas gloves are not recommended for the same reason. The worker should not apply pesticides at home. Their clothing should be double washed separately from the family wash. Their personal vehicle should be washed and vacuumed. Their work vehicle should also be washed and vacuumed. Remove the worker from exposure for 1 wk and retest. If after the retest, RBC AChE depression is the same or greater the worker should see a physician.

3. Neither blood esterase is depressed, but the worker complains of any or a combination of fatigue, weakness, dizziness, headaches, nausea, intestinal problems, diarrhea, sweating, blurred vision, and vomiting.

A pesticide worker complaining of any of these symptoms should be removed from exposure. Send this person to a physician. Monitor esterase levels immediately. Counsel the worker and take the hygiene precautions as in #2. The worker is complaining now. Blood tests normally take a day or two for results to be communicated, and an exposure could have taken place in the meantime. Symptoms are normally accompanied by esterase depression, so a retest will help separate pesticide exposure from another illness such as the flu.

4. Both esterase activities are depressed 20%–30%, but the worker feels great and is ready to spray. In fact, the worker insists on going to work.

Remove the worker from exposure. It has been shown time after time in humans that symptoms disappear, but blood esterases remain depressed. It is speculated this is because the nervous system enzyme either recovers more quickly or is produced more rapidly than the blood enzymes. The OPs are cumulative poisons. This worker is headed for an acute illness with continued exposure. Counsel the worker and take the hygiene precautions as in #2.

5. During the baseline determinations, plasma BChE is either 30%–60% low or is very low, around 2%. RBC AChE is normal.

Low plasma BChE (compared to the population range) is usually genetic, but a 30%–60% reduction might result from a pathological condition. The attending physician can determine which situation is correct. If low plasma BChE is caused by a health problem, the worker must not apply pesticides. If low plasma BChE is genetic, the worker should be counseled about this fact. People with low plasma BChE do not metabolize some anesthetics rapidly, so this information will be useful for a physician in the future. This worker can apply OPs, but more reliance must be placed on RBC AChE, so reduce the percentage removal criteria by 5%, e.g., 30% to 25%. If you have chosen 20% reduction in RBC AChE as a removal criterion, this worker should not apply OPs because normal laboratory variation may be 15%. There is debate whether a person with genetically very low (< 2%) plasma BChE should apply OPs. People with 2% or less of 'normal' BChE live normal lives, and no other health problem is associated with this genetic condition. A monitoring tool is absent, so the recommendation is that a 'silent' plasma BChE worker should not apply OPs.

Once esterases are depressed, the worker should be removed from exposure until the level(s) return to the preexposure range. The time for this to occur depends on the level of depression. For plasma BChE, this could be 5–30 d, and up to 90 d for RBC AChE.

It will definitely help you and the worker if the testing procedures and criteria are fully explained. The OP pesticides *depress* RBC AChE and plasma BChE. However, workers may be told by their physician that they have been exposed when their esterase values are above the population normal. Any person may be outside the population normal, either high or low, and this may simply be that person's normal value. However, if esterase values rise above the worker's baseline values, redo the baselines because the worker may have been exposed when the baselines were determined.

Several events occur when a blood esterase monitoring program is part of a safety program. The morale of the workforce increases. Workers have tangible evidence that they are important to the overall operation and that their health and person are important to the supervisor. They have an opportunity to participate in a health program and to see how their work practices affects them personally. For the supervisor, one piece of evidence that other safety practices are working is in hand. Careless workers can be singled out for counseling and training. Insurers can be approached for a reduction in rates, and insurers will reduce rates when presented with an overall safety program.

A blood monitoring program works as one facet of a comprehensive pesticide safety program. There is no substitute for good work practices. Good work practices should be a condition of employment, not voluntary. Issue clean clothing on a daily basis, require showers after spraying, and provide handwashing facilities, clean protective gear, clean equipment, and thorough training. Many of these requirements are part of the new worker protection standards (USEPA 1992). Careful supervision of a comprehensive safety program will prevent accidents, minimize pesticide exposure, maximize health and productivity, and reduce liability.

XV. The Future: Physiologically Based Pharmacokinetic Models

Physiologically based pharmacokinetic/pharmacodynamic (PBPK/PD) models are needed to predict the fate (absorption, activation, detoxification, inhibition of AChE/BChE, and elimination) of OPs in animals and extrapolate the results from animals to humans for risk assessments. PBPK models were recently developed for DEF, parathion, malathion, and isofenphos by Gearhart et al. (1990), Sultatos (1990), Dong et al. (1996), and Knaak et al. (1996), respectively. Gearhart et al. (1990) modeled the inhibition of AChE and BChE by DFP and its hydrolysis to HF and *O,O*-diisopropyl phosphate. Sultatos (1990) modeled the i.v. absorption of parathion in the mouse and oxidative metabolism (activation and detoxification) to paraoxon and *p*-nitrophenol. Dong et al. (1996) modeled the human absorption/metabolism/elimination of a topically applied dose of malathion. Knaak et al. (1996) modeled the dermal absorption of isofenphos in the rat, oxidative/hydrolytic metabolism in the liver, and inhibition of AChE and BChE in brain and blood by the toxic metabolite, des *N*-isopropyl isofenphos oxon. The metabolic pathway for isofenphos is given in Fig. 3 for the rat, guinea pig, and dog. In the rat, AChE and BChE inhibition are dependent upon the V_{max}, K_m values (Knaak et al. 1993b) for the P-450-catalyzed formation of oxon, hydrolysis of oxon by 'A' esterases and the bimolecular rate constants (k_i) for the phosphorylation of ChEs by the toxic oxon. Rats absorbed 47% of a topical dose of isofenphos over 168 hr (Knaak et al. 1996), while humans absorbed only 3.6% over 72 hr (Wester et al. 1992). Evaporative losses (*in vivo*) of isofenphos accounted for 45% in the rat study and >90% in the human study. A PBPK model is needed for chlorpyrifos. This OP pesticide is used extensively for insect control around the home where children are likely to contact residues of this OP insecticide.

The previous sections have presented the background and considerations involved in a worker esterase monitoring program. There are many deficiencies in the program. One of the best ways to see these weaknesses and to correct them is through physiologically based/pharmacodynamic (PBPK/PBPD) models (Knaak et al. 1993a, 1994, 1996). The strength of these models is that every process involved in the inhibition of an esterase is considered: exposure route, metabolism, enzyme induction, activation, excretion, organ blood flow, metabolite disposition, detoxification routes, and so forth. For instance, a percutaneous PBPK model of OP pesticide exposure requires pharmacokinetic evaporation data from skin, skin permeability, metabolic pathways and their kinetic parameters, partition coefficients, and elimination (excretion) data. For esterase inhibitors, the human phenotype for BChE may also be required. PBPK models are also compound specific. This specificity is their disadvantage. The expense of gathering the PBPK model data for one compound is very high. For many compounds the cost is prohibitive. Nonetheless the PBPK model approach forces consideration of the biochemistry and physiology of the whole organism as opposed to the narrow view of only esterases or another target system.

XVI. Discussion

During the past 15 yr, USEPA reregistration of pesticide products containing the same active ingredients (OP) has resulted in a reduction (by cancellation of registration or other means) in the total number (~40) of active OPs available for use in pesticide products in the U.S. The reasons for cancellation are in part, but not entirely related to the failure of registrants to provide the USEPA with adequate technical data (i.e, toxicology, environmental fate/exposure, ecological effects and product/residue chemistry) to support continued registration and use of their active ingredients or pesticide products. Neurotoxicity (neurochemistry/neuropathology, physiology, behavior, sensory/motor, electrophysiology, and learning/memory behavior) is of principal concern to the USEPA when reviewing dose–response studies involving ChE inhibitors (Raffaele and Rees 1990). For example, Sheets et al. (1997) conducted neurotoxicity studies on six OPs (sulprofos, disulfoton, azinphosmethyl, methamidophos, trichlorfon, and tebupirimiphos) according to the USEPA (1991) requirements. All treatment-related neurobehavioral findings were ascribed to cholinergic toxicity occurring only at dietary levels that produced more than 20% inhibition of plasma, RBC, and brain ChE activity. The functional observation battery (FOB) and motor activity (MA) findings did not alter the conclusions and generally did not reduce the no-observed-effect level (NOEL) relative to detailed clinical observations. Study supports a general standard of more than 20% inhibition of brain ChE activity for cholinergic neurotoxicity.

A detailed analysis of the technical data must be made by the USEPA as it relates to the pesticide, route and length of exposure, dose, dose rate, animal (species, sex, and age), biotransformation, and response (ChE inhibition, NOEL) for establishing RfDs, setting food tolerances, and monitoring and controlling worker exposure. Our review provides information on currently registered OPs (and other ChE inhibitors), metabolic pathway data, metabolic enzymes, neurotoxicity, tissue and structural information on ChEs, and kinetics of ChE-catalyzed reactions and inhibition (phosphorylation) by OPs relative to blood AChE/BChE monitoring and the development of PBPK/PD models.

Well-constructed and validated PBPK/PD models may be used to obtain human equivalent concentrations ($NOEL_{[HEC]}$) from animal no-effect-levels (NOEL) by extrapolation of pesticide (oxon) peak height or area-under-curve (AUC) in animal tissues to humans or by the extrapolation of AChE inhibition values from animals to humans. $NOEL_{[HEC]}$ are used by USEPA to determine RfDs and by the Agency for Toxic Substances and Disease Registry (ATSDR) for calculating minimal risk levels (MRLs). Workplace environmental exposure standards (inhalation with skin notation) were recently adopted by NIOSH/OSHA for parathion, azinphosmethyl, carbaryl, demeton, dichlorvos (DDVP), EPN, malathion, phosdrin, fenchlorphos (Ronnel®) and TEPP (NIOSH 1990).

These methods are incomplete or under development, so the monitoring of blood AChE and BChE activity and alkyl phosphates in urine of workers exposed to pesticides during mixing loading and application continues to be the best way of preventing illnesses. The effects of OPs on human behavior were investigated in mixer-loader applicators in California during the 1970s (Knaak 1998, personal communication). The California studies were conducted for purposes of obtaining information on direct effects on the central and peripheral nervous system. Unfortunately, behavioral changes in mixer-loader applicators did not provide useful information as exposed workers were capable of compensating for small neurological deficits and the work was largely discontinued despite its early popularity. Eyer (1995) reviewed the available literature dealing with neuropsychological changes after exposure to OP insecticides. Long-term toxic effects affecting behavior as well as mental and visual functions occasionally occur as the result of repeated acute, clinically significant intoxications. No increased risk of delayed or permanent neuropsychopathological effects were connected to asymptomatic exposure to OPs. Acute exposures result in subchronic neurological sequelae, which include a myopathy brought about by excess ACh and in some cases OPIDN caused by the inhibition of neuropathy target esterase.

The cancellation of OP products containing phosalone, dialifor, parathion, and phosdrin has substantially decreased the hazards involved in pesticide use. Regardless of this fact, worker safety regulations on closed mixing and loading systems, reentry intervals, medical supervision, illness reporting, and blood AChE monitoring should remain in effect. According to Krieger (1995), additional information is needed regarding human exposure to OPs in the workplace and home. Field poisoning incidents involving phosalone on grapes (O'Malley and McCurdy 1990) are now believed to be the result of exposure to leaf residues ($\mu g/cm^2$) above those reported in field measurements (Krieger 1998, personal communication) or perhaps they are the result of human sensitivity to phosalone and phosalone oxon greater than the test animals.

In our monitoring recommendations, we have presented a practical method which considers, but does not require, the information required to construct a PBPK/PD model. Our monitoring recommendations assume that inhibition of AChE/BChE in blood is a reflection of changes occurring in the central nervous system. However, nervous system symptoms can disappear well before blood enzymes recover and blood enzymes can be 100% inhibited with few or no symptoms (see Grob references; Kusic et al. 1991). In the case of heptylphysostigmine (a carbamate), the reverse is true. RBC AChE recovers in 24 hr while brain AChE remains inhibited (Moriearty and Becker 1993), indicating that RBC AChE and brain AChE react differently with some compounds. For BChE, one or two amino acid changes (which occurs in various human genotypes) can render BChE resistant to inhibitors and/or to reactivation (Neville et al. 1990).

We debated whether inhibition of either RBC AChE or plasma BChE should be allowed based on differential RBC AChE vs BChE inhibition. If brain enzyme is inhibited differentially compared to blood enzymes, monitoring a blood enzyme would give a false sense of security. As differential enzyme inhibition can occur with one chemical, the potential for differential enzyme inhibition exists for any chemical.

Our recommendations differ from current practices in California. California regulations allow 30% depression of RBC AChE and 40% depression of BChE before removal of a worker from the work place (Wilson et al. 1998). From the previous discussion, about 15% variation can be expected in measurement of either enzyme under tightly controlled laboratory conditions. Under clinical use RBC AChE activities were overestimated in inhibited samples (Wilson et al. 1998). It is easy to see that the inaccuracies in measurement of either blood enzyme could miss an 80%–100% depression. Our recommendations allow about twice the 15% measurement variation in a well-run laboratory. With a decrease in measurement variation, the 25%–30% depression removal criteria we recommend could be lowered another 5%. This would be close to the best scientific effort to determine zero AChE and BChE depression.

Summary

The organophosphorus pesticides of this review were discovered in 1936 during the search for a replacement for nicotine for cockroach control. The basic biochemical characteristics of RBC AChE and BChE were determined in the 1940s. The mechanism of inhibition of both enzymes and other serine esterases was known in the 1940s and, in general, defined in the 1950s. In 1949, the death of a parathion mixer-loader dictated blood enzyme monitoring to prevent acute illness from organophosphorus pesticide intoxication. However, many of the chemical and biochemical steps for serine enzyme inhibition by OP compounds remain unknown today. The possible mechanisms of this inhibition are presented kinetically beginning with simple (by comparison) Michaelis–Menten substrate enzyme interaction kinetics. As complicated as the inhibition kinetics appear here, PBPK model kinetics will be more complex.

The determination of inter- and intraindividual variation in RBC ChE and BChE was recognized early as critical knowledge for a blood esterase monitoring program. Because of the relatively constant production of RBCs, variation in RBC AChE was determined by about 1970. The source of plasma (or serum) BChE was shown to be the liver in the 1960s with the change in BChE phenotype to the donor in liver transplant patients. BChE activity was more variable than RBC AChE, and only in the 1990s have BChE individual variation questions been answered.

We have reviewed the chemistry, metabolism, and toxicity of organophosphorus insecticides along with their inhibitory action toward tissue acetyl- and butyrylcholinesterases. On the basis of the review, a monitoring program for individuals mixing-loading and applying OP pesticides for commercial applicators was recommended.

Approximately 41 OPs are currently registered for use by USEPA in the United States. Under agricultural working conditions, OPs primarily are absorbed through the skin. Liver P-450 isozymes catalyze the desulfurization of phosphorothioates and phosphorodithioates (e.g., parathion and azinphosmethyl, respectively) to the more toxic oxons (P=O(S to O)). In some cases, P-450 isozymes catalyze the oxidative cleavage of P–O–aryl bonds (e.g., parathion, methyl parathion, fenitrothion, and diazinon) to form inactive water-soluble alkyl phosphates and aryl leaving groups that are readily conjugated with glucuronic or sulfuric acids and excreted. In addition to the P-450 isozymes, mammalian tissues contain ('A' and 'B') esterases capable of reacting with OPs to produce hydrolysis products or phosphorylated enzymes. 'A'-esterases hydrolyze OPs (i.e, oxons), while 'B'-esterases with serine at the active center are inhibited by OPs. OPs possessing carboxylesters, such as malathion and isofenphos, are hydrolyzed by the direct action of 'B'-esterases (i.e, carboxylesterase, CaE). Metabolic pathways shown for isofenphos, parathion, and malathion define the order in which these reactions occur, while Michaelis–Menten kinetics define reaction parameters (V_{max}, K_m) for the enzymes and substrates involved, and rates of inhibition of 'B'-esterases (k_is, bimolecular rate constants) by OPs and their oxons.

OPs exert their insecticidal action by their ability to inhibit AChE at the cholinergic synapse, resulting in the accumulation of acetylcholine. The extent to which AChE or other 'B'-esterases are inhibited in workers is dependent upon the rate the OP pesticide is activated (i.e., oxon formation), metabolized to nontoxic products by tissue enzymes, its affinity for AChE and other 'B'-esterases, and esterase concentrations in tissues. Rapid recovery of OP BChE inhibition may be related to reactivation of inhibited forms. AChE, BChE, and CaE appear to function *in vivo* as scavengers, protecting workers against the inhibition of AChE at synapses. Species sensitivity to OPs varies widely and results in part from binding affinities (K_a) and rates of phosphorylation (k_p) rather than rates of activation and detoxification. Neurotoxicity, once thought to be entirely caused by the inhibition of AChE and regeneration, may involve a second messenger as well as a transmitter other than the cholinergic system.

The relationship between exposure, percutaneous absorption, metabolism, and cholinesterase inhibition is not well understood. Physiologically based pharmacokinetic/pharmacodynamic models are currently being developed to enable researchers to extrapolate animal exposure data to workers to establish workplace exposure standards for OPs.

The activity of RBC AChE in blood samples shows about 11% interindividual variation for men and women. Plasma activity varies about 20% in men and women, with individual variation in plasma BChE concentration and activity estimated to be 6.4% and 6.2%, respectively. Lack of standardized esterase assays and quality controls in clinical laboratories that measure blood AChE and BChE activity continues to hamper efforts to monitor OP exposure in humans. In fact, the failure to standardize AChE and BChE clinical assays based on known biochemical enzyme characteristics is inexcusable. Human health conditions such as liver disease, anemia, and acute infectious diseases, generic variants of human serum plasma BChE, and diets containing alkaloid ChE inhibitors exert a negative effect on plasma BChE activity and may make it difficult to interpret the significance of clinical values.

The purposes of this review were to present the scientific details that result in a human blood monitoring program for pesticide exposure and to suggest a monitoring program. The scientific details are complex, but those details suggest the confidence we should have in the final monitoring program. We know with certainty that individuals vary for genetic and sometimes for environmental reasons in their ChE activities. We know with certainty that genetic variation in the structure of ChE can lead to resistance to inhibition by some inhibitors. We know with certainty that differential inhibition of tissue ChEs can occur. We do not know all the inhibitors that differentially inhibit ChE. We do not know with certainty how genetic groups react to various inhibitors because not all inhibitors or all genetic groups have been tested.

Based on these data, our recommendations for a blood esterase monitoring program include but are not limited to (1) obtaining a complete medical history for each employee and a physical examination; (2) obtaining the services of a competent clinical laboratory licensed to run blood cholinesterase tests; (3) measuring a worker's blood esterase activity (at least two analyses) before employment; (4) setting up a monitoring program based on duration and level of exposure to OPs; 5) removing employee from work if a 25%–30% reduction in RBC AChE occurs; (6) determining cause of exposure and making adjustments to prevent repeated exposure; and (7) allowing employee to return to work when RBC AChE activity returns to preexposure values.

In a general sense, we know that following these recommendations leads to the prevention of short-term illness from OP pesticide exposure. Any specific individual could be the exception to the rule. The exceptions hardly have been studied and, consequently, we urge care and compassion in the conduct of any OP pesticide exposure monitoring program.

References

Aldridge WN (1950) Some properties of specific cholinesterase with particular reference to the mechanism of inhibition by diethyl p-nitrophenyl thiophosphate (E 605) and analogues. Biochem J 46:451–460.

Aldridge WN (1953a) Serum esterases. I. Two types of esterase (A and B) hydrolyzing p-nitrophenyl acetate, propionate, and butyrate, and a method for their determination. Biochem J 53:110–117.

Aldridge WN (1953b) Differentiation of true and pseudo cholinesterase by organophosphorus compounds. Biochem J 53:62–67.

Aldridge WN (1954a) Tricresyl phosphates and cholinesterase. Biochem J 56:185–189.

Aldridge WN (1954b) Anticholinesterases. Inhibition of cholinesterase by organophosphorus compounds and reversal of this reaction. Mechanism involved. Chem Ind 473–476.

Aldridge WN, Davison AN (1953) Mechanism of inhibition of cholinesterases by organophosphorus compounds. Biochem J 55:763–765.

Aldridge WN, Reiner E (1972) Enzyme inhibitors as substrates. In: Neuberger A, Tatum EL (eds) North-Holland Research Monographs, Frontiers of Biology, Vol. 26. North-Hollands, London, p 236.

Alles GA, Hawes RC (1940) Cholinesterases in the blood of man. J Biol Chem 133:375–390.

Alozie SO, Sharma RP, Salunkhe DK (1978) Inhibition of rat cholinesterase isoenzymes in vitro and in vivo by the potato alkaloid, α-chaconine. J Food Biochem 2:259–276.

Areekul S, Srichairat S, Kirdudom P (1981) Serum and red cell cholinesterase activity in people exposed to organophosphate insecticides. Southeast Asian J Trop Med Public Health 12:94–98.

Arpagaus M, Chatonnet A, Masson P, Newton M, Vaughan TA, Bartels CF, Nogueira CP, La Du BN, Lockridge O (1991) Use of the polymerase chain reaction for homology probing of butyrylcholinesterase from several vertebrates. J Biol Chem 266:6966–6974.

Atack JR, Perry EK, Bonham JR, Perry RH (1987) Molecular forms of acetylcholinesterase and butyrylcholinesterase in human plasma and cerebrospinal fluid. J Neurochem 48:1845–1850.

Atack JR, Yu QS, Soncrant TT, Brossi A, Rapoport SI (1989) Comparative inhibitory effects of various physostigmine analogs against acetyl- and butyrylcholinesterases. J Pharmacol Exp Ther 249:194–202.

Atkins EL, Rubins CH, Olsoni DR, Jackson RJ (1998) Rapid assessment of organophosphate-reduced cholinesterase depression: a comparison of laboratory and field kit methods to detect human exposure to organophosphates. Appl Occup Environ Hyg 265–268.

Augustinsson KB (1948) Cholinesterase: a study in comparative enzymology. Acta Physiol Scand Suppl 15:1–182.

Augustinsson KB (1949) Substrate concentration and specificity of choline ester-splitting enzymes. Arch Biochem 23:111–126.

Augustinsson KB (1955) The normal variation of human blood cholinesterase activity. Acta Physiol Scand 35:40–52.

Augustinsson KB (1960) Butyryl- and propionylcholinesterases and related types of eserine-sensitive esterases. In: Boyer PD, Lardy H, Myrbdck (eds) The Enzymes, 2nd Ed., Vol. 4. Academic Press, New York, pp 521–540.

Bell JU, Van Petten GR, Taylor PJ, Aiken MJ (1979) The inhibition and reactivation of human maternal and fetal plasma cholinesterase following exposure to the organophosphate, dichlorvos. Life Sci 24:247–254.

Benjamini E, Metcalf RL, Fukuto TR (1959a) The chemistry and mode of action of the insecticide O,O-diethyl O-p-methyl sulfinyl phenyl phosphorothionate and its analogues. J Econ Entomol 52:94–98.

Benjamini E, Metcalf RL, Fukuto TR (1959b) Contact and systemic insecticidal properties of O,O-diethyl O-p-methyl sulfinyl phenyl phosphorothionate and its analogues. J Econ Entomol 52:99–102.

Berman HA (1995) Reaction of acetylcholinesterase with organophosphonates. In: Quinn DM et al. (eds) The Enzymes of the Cholinesterase Family. Plenum Press, New York, pp 177–182.

Berman HA, Decker MM (1989) Chiral nature of covalent methylphosphonyl conjugates of acetylcholinesterase. J Biol Chem 264:3951–3956.

Berman HA, Leonard K (1989) Chiral reactions of acetylcholinesterase probed with enantiomeric methylphosphorothioates. J Biol Chem 264:3942–3950.

Bernsohn J, Barron KD, Hess A (1961) Cholinesterases in serum as demonstrated by starch gel electrophoresis. Proc Soc Exp Biol Med 108:71–73.

Bogusz M (1968) Influence of insecticides on the activity of some enzymes contained in human serum. Clin Chim Acta 19:367–369.

Bonham JR, Gowenlock AH, Timothy JAD (1981) Acetylcholinesterase and butyrylcholinesterase measurement in the pre-natal detection of neural tube defects and other fetal malformations. Clin Chim Acta 115:163–170.

Bowman JS, Casida JE (1957) Metabolism of the systemic insecticide O,O-diethyl S-ethylthiomethyl phosphorodithioate (Thimet) in plants. J Agric Food Chem 5:192-197.

Brauer RW (1948) Inhibition of the cholinesterase activity of human blood plasma and ereythrocyte stromata by alkylated phosphorus compounds. J Pharmacol Exp Ther 92:162–172.

Brock A (1990) Immunoreactive plasma cholinesterase (EC 3.1.1.8) substance concentration, compared with cholinesterase activity concentration and albumin: inter- and intra-individual variations in a healthy population group. J Clin Chem Clin Biochem 28:851–856.

Brock A (1991) Inter and intraindividual variations in plasma cholinesterase activity and substance concentration in employees of an organophosphorous insecticide factory. Br J Ind Med 48:562–567.

Brock A, Brock V (1993) Factors affecting inter-individual variation in human plasma cholinesterase activity: body weight, height, sex, genetic polymorphism and age. Arch Environ Contam Toxicol 24:93–99.

Bull DL (1970) Metabolism of organophosphorus insecticides. Presented at the Annual Meeting of the Entomological Society of America, Miami.

Bull DL, Linquist DA (1964) Metabolism of 3-hydroxy-N,N-dimethyl-crotonamide dimethyl phosphate by cotton plants, insects and rats. J Agric Food Chem 12:310–317.

Bull DL, Linquist DA (1966) Metabolism of 3-hydroxy-N-methyl-*cis*-crotonamide dimethyl phosphate (azodrin). J Agric Food Chem 14:105–109.

Bull DL, Lindquist DA, Grabbe RR (1967) Comparative fate of the geometric isomers of phosphamidon in plants and animals. J Econ Entomol 60:332–341.

Callaway S, Davies DR, Rutland JP (1951) Blood cholinesterase levels and range of personal variation in a healthy adult population. Br Med J 2:812–816.

Carter MK, Maddux B (1974) Interaction of dichlorvos and anticholinesterases on the *in vitro* inhibition of human blood cholinesterases. Toxicol Appl Pharmacol 27:456–463.

Casida JE (1956) Metabolism of organophosphorus insecticides in relation to their antiesterase activity, stability, and residual properties. J Agrc Food Chem 4:772–785.

Chambers JE (1992) The role of target site activation of phosphorothionates in acute toxicity. In: Chambers JE, Levi PE (eds) Organophosphates Chemistry, Fate, and Effects. Academic Press, New York, pp 229–239.

Chambers JE, Ma T, Boone JS, Chambers HW (1994) Role of detoxification pathways in acute toxicity levels of phosphorothionate insecticides in the rat. Life Sci 54:1357–1364.

Chatonnet A, Lockridge O (1989) Comparison of butyrylcholinesterase and acetylcholinesterase. Biochem J 260:625–634.

Clemmons GP, Menzer RE (1968) Oxidative metabolism of phosphamidon in rats and a goat. J Agric Food Chem 16:312–318.

Cohen JA, Warringa MGPJ (1957) Purification and properties of dialkyl fluorophosphatase. Biochim Biophys Acta 26:29–39.

Cohen JA, Oosterbaan RA, Warringa MGPA (1955) Turnover number of ali-esterase, pseudo- and true cholinesterase and the combination of these enzymes with diisopropyl fluorophosphate (DFP). Biochim Biophys Acta 18:228–235.

Cohen SD, Williams RA, Killinger JM, Fredenthal RI (1985) Comparative sensitivity of bovine and rodent acetylcholinesterase to in vitro inhibition by organophosphate insecticides. Toxicol Appl Pharmacol 81:452–459.

Comroe JH Jr, Todd J, Gammon GD, Leopold IH, Koelle GB, Bodansky O, Gilman A (1946a) The effect of di-isopropyl- fluorophosphate (DFP) upon patients with myasthenia gravis. Am J Med Sci 212:641–651.

Comroe JH Jr, Todd J, Koelle GB (1946b) The pharmacology of di-isopropyl fluorophosphate (DFP) in man. J Pharmacol Exp Ther 87:281–290.

Cook JW, Yip G (1958) Malathionase. II. Identity of a malathion metabolite. J Assoc Off Agric Chem 41:407–411.

Cook JW, Blake JK, Yip G, Williams M (1958) Malathionase. I. Activity and inhibition. J Assoc Off Agric Chem 41:399–407.

Dauterman WC (1971) Biological and non-biological modification of organophosphorus compounds. Bull World Health Org 44:133–150.

Dauterman WC, Viado GB, Casida JE, O'Brien RD (1960) Persistence of dimethoate and metabolites following foliar application to plants. J Agric Food Chem 8:115–119.

Dawson RM (1990) Reversibility of the inhibition of acetylcholinesterase by tacrine. Neurosci Lett 118:85–87.

Dean RA, Christian CD, Barry Sample RH, Bosron WF (1991) Human liver cocaine esterases: ethanol-mediated formation of ethylcocaine. Fed Am Soc Exp/Biol Monogr 5:2735–2739.

de Jong LPA, Van Dijk C, Benschop HP (1989) Stereoselective hydrolysis of soman and other chiral organophosphates by mammalian phosphorylphosphatases. In: Reiner E, Aldridge WN, Hoskin FCG, Horwood E (eds) Enzymes Hydrolysing Organophosphorus Compounds. Limited, Chichester, England, pp 65–78.

Domino EF (1988) Galanthamine: another look at an old cholinesterase inhibitor. In: Giacobini E, Becker R (eds) Current Research in Alzheimer Therapy. Taylor and Francis, New York, pp 295–303.

Dong MH, Ross JH, Thongsinthusak T, Krieger RI (1996) Use of spot urine sample results in physiologically based pharmacokinetic modeling of absorbed malathion doses. In: Blancato JN, Brown RN, Dary CC, Saleh MA (eds) Biomarkers for Agrochemicals and Toxic Substances: Applications and Risk Assessment. ACS Symposium Series 643. American Chemical Society, Washington, DC.

Donninger C, Hutson HD, Pickering BA (1967) Oxidative cleavage of phosphoric acid triesters to diesters. Biochem J 102:26–27.

Douch PGC, Hook CER, Smith JN (1968) Metabolism of Folithion (O,O-dimethyl O-(4-nitro-m-tolyl) phosphorothioate). Aust J Pharm 49:S70-S71.

Drevenkar V, Radic Z, Vasilic Z, Reiner E (1991) Dialkylphosphorus metabolites in the urine and activities of esterases in the serum as biochemical indices for human absorption of organophosphorus pesticides. Arch Environ Contam Toxicol 20:417–422.

Duncan RC, Griffith J, Konefal J (1986) Comparison of plasma cholinesterase depression among workers occupationally exposed to organophosphorus pesticides as reported by various studies. J Toxicol Environ Health 18:1–11.

Eckerson HW, Oseroff A, Lockridge O, La Du BN (1983) Immunological comparison of the usual and atypical human serum cholinesterase phenotypes. Biochem Genet 21:93–108.

Ecobichon DJ, Stephens DS (1972) Perinatal development of human blood esterases. Clin Pharmacol Ther 14:41–47.

El Bashir S, Oppenoorth FJ (1969) Microsomal oxidations of organophosphate insecticides in some resistance strains of houseflies. Nature (Lond) 223:210–211.

Ellman GL, Courtney KD, Andres V Jr, Featherstone RM (1961) A new and rapid colorimetric determination of acetylcholinesterase activity. Biochem Pharmacol 7:88–95.

Enslein K, Gombar VK, Shapero D, Blake BW (1998) Prediction of rat oral LD_{50} values of organophosphates by QSAR equations. Topkat Health Designs, Rochester, NY.

Eto M (1974) Organophosphorus Pesticides: Organic and Biological Chemistry. CRC Press, Boca Raton, FL, p 164.

Eyer P (1995) Neuropsychopathological changes by organophosphorus compounds: a review. Hum Exp Toxicol 14:857–864.

Frawley JP, Hagan EC, Fitzhugh OG (1952) A comparative pharmacological and toxicological study of organic phosphate—anticholinesterase compounds. J Pharmacol Exp Ther 105:156–164.

Freedman AM, Willis A, Himwich HE (1949) Correlation between signs of toxicity and cholinesterase level of brain and blood during recovery from di-isopropyl fluorophosphate (DFP) poisoning. Am J Physiol 57:80–87.

Fukami J, Shishido T (1966) Nature of the soluble, glutathione-dependent enzyme system active in cleavage of methyl parathion to desmethyl parathion. J Econ Entomol 59:1338–1346.

Fukunaga K (1967) The in vitro metabolism of organophosphorus insecticides by tissue homogenates from mammals and insect. In: U. S.–Japan Seminar, Experimental Approaches to Pesticide Metabolism, Degradation, and Mode of Action, pp 197–207.

Fukunaga K, Fukami J, Shishido T (1969) The *in vitro* metabolism of organophosphorus insecticides by tissue homogenates from mammal and insect. Residue Rev 25:223–250.

Fukuto TR (1971) Relationship between the structure of organophosphorus compounds and their activity as acetylcholinesterase inhibitors. Bull World Health Org 44:31–42.

Fukuto TR, Metcalf RL (1956) Structure and insecticidal activity of some diethyl substituted phenyl phosphates. J Agric Food Chem 4:930–935.

Fukuto TR, Metcalf RL, March RB, Maxon MG (1955) Chemical behavior of systox isomers in biological systems. J Econ Entomol 48:347–354.

Fukuto TR, Wolf JP, Mecalf RL, March RB (1956) Identification of the sulfoxide and sulfone plant metabolites of the thiol isomer of systox. J Econ Entomol 49:147–151.

Gage JC (1953) A cholinesterase inhibitor derived from O, O-diethyl O-p-nitrophenyl. Biochem J 54:426–430.

Gage JC (1955) Blood cholinesterase values in early diagnosis of excessive exposure to phosphorus insecticides. Br Med J 1:1370–1372.

Gage JC (1967) The significance of blood cholinesterase activity measurements. Residue Rev 18:159–173.

Ganelin RS (1964) Measurement of cholinesterase activity of the blood. Ariz Med 21:710–714.

Gearhart JM, Jepson GW, Clewell HJ, Andersen ME, Connolly RB (1990) Physiologically based pharmacokinetic and pharmacodynamic model for the inhibition of acetylcholinesterase by diisopropylfluorophosphate. Toxicol Appl Pharmacol. 106:295–310.

Geller I, Stebbins WC (1979) Introduction and overview. In: Geller I, Stebbins WC, Wayner MJ (eds) Proceedings of the Workshop on Test Methods for Definition of Effects of Toxic Substances on Behavior and Neuromotor Function. Neurobehav Toxicol 1:7.

Geller I, Stebbins WC, Wayner MJ (1979) Proceedings of the Workshop on Test Methods for Definition of Effects of Toxic Substances on Behavior and Neuromotor Function. Neurobehavioral Toxicology 1. Ankho International, Fayetteville, NY 225 pp

Glick D (1937) Properties of choline esterase in human serum. Biochem J 31:521–525.

Gnatt A, Ginzberg D, Lieman-Hurwitz J, Zamir R, Zakut H, Soreq H (1991) Human acetylcholinesterase and butyrylcholinesterase are encoded by two distinct genes. Cell Mol Neurobiol 11:91–104.

Goldberg ME, Johnson HE, Knaak JB (1965) Inhibition of discrete avoidance behavior by three anticholinesterase agents. Psychopharmacologia 7:72–76.

Goldberg ME, Johnson HE, Knaak JB, Smyth HF (1963) Psychopharmacological effects of reversible cholinesterase inhibition induced by N-methyl 3-isopropyl phenyl carbamate (compound 10854). J Pharmacol Exp Ther 141:244–252.

Goldstein A (1951) Properties and behavior of purified human plasma cholinesterase. III. Competitive inhibition by prostigmine and other alkaloids with special reference to differences in kinetic behavior. Arch Biochem Biophys 34:169–188.

Griffiths JT, Sterns CR, Thompson WL (1951) Parathion hazards encountered spraying citrus in Florida. J Econ Entomol 44:160–163.

Grob D (1950) The anticholinesterase activity *in vitro* of the insecticide parathion (*p*-nitrophenyl diethyl thionophosphate). Bull Johns Hopkins Hosp 87:95–105.

Grob D (1956) The manifestations and treatment of poisoning due to nerve gas and other organic phosphate anticholinesterase compounds. Arch Intern Med 98:221–239.

Grob D, Harvey AM (1949) Observations on the effects of tetraethyl pyrophosphate (TEPP) in man, and on its use in the treatment of myasthenia gravis. Bull Johns Hopkins Hosp 84:532–566.

Grob D, Harvey JC (1958) Effects in man of the anticholinesterase compound sarin (isopropyl methyl phosphonofluoridate). J Chem Invest 37:350–368.

Grob D, Lilienthal JL Jr, Harvey AM, Jones BF (1947a) The administration of di-isopropyl fluorophosphate (DFP) to man. I. Effect on plasma and erythrocyte cholinesterase; general systemic effects; use in study of hepatic function and erythropoiesis; and some properties of plasma cholinesterase. Bull Johns Hopkins Hosp 81:217–245.

Grob D, Lilienthal JL Jr, Harvey AM (1947b) The administration of di-isopropyl fluorophosphate (DFP) to man. II. Effect on intestinal motility and use in the treatment of abdominal distention. Bull Johns Hopkins Hosp 81:245–256.

Grob D, Harvey AM, Langworthy OR, Lilienthal JL Jr (1947c) The administration of di-isopropyl fluorophosphate (DFP) to man. III. Effect on the central nervous system with special reference to the electrical activity of the brain. Bull Johns Hopkins Hosp 81:257–292.

Grob D, Garlick WL, Harvey AM (1950) The toxic effects in man of the anticholinesterase insecticide parathion (*p*-nitrophenyl diethyl thionophosphate). Bull Johns Hopkins Hosp 87:106–129.

Hada T, Ohue T, Imanishi H, Nakaoka H, Hirosaki A, Shimomura S, Fujikura M, Matsuda Y, Yamamoto T, Amuro Y, Higashino K (1990) Discrimination of liver cirrhosis from chronic hepatitis by analysis of serum cholinesterase isozymes using affinity electrophoresis with concanavalin A or wheat germ agglutinin. Gastroenterol Jpn 25:715–719.

Hansen ME, Wilson BW (1999) Oxime reactivation of RBC acetylcholinesterases for biomonitoring. Arch Environ Contam Toxicol 37:283–289.

Harris MH, Harris RS (1944) Effect *in vitro* of curare alkaloids and crude curare preparations on "true" and pseudo-cholinesterase activity. Proc Soc Exp Biol Med 56:223–225.

Harris H, Robson EB (1963) Fractionation of human serum cholinesterase components by gel filtration. Biochim Biophys Acta 73:649–652.

Harris H, Whittaker M (1959) Differential response of human serum cholinesterase types to an inhibitor in potato. Nature (Lond) 183:1808–1809.

Harris H, Whittaker M (1962) Differential inhibition of the serum cholinesterase phenotypes by solanine and solanidine. Ann Hum Genet 26:73–76.

Hart GJ, O'Brien RD (1973) Recording spectrophometric method for determination of dissociation and phosphorylation constants for the inhibition of acetylcholinesterase by
organophosphates in the presence of substrate. Biochem 12:2940–2945.

Hartwell WV, Hayes GR Jr, Funckes AJ (1964) Respiratory exposure of volunteers to parathion. Arch Environ Health 8:820–825.

Harvey AM, Lilienthal JL Jr, Grob D, Jones BF, Talbot SA (1947) The administration of
di-isopropyl fluorophosphate to man. IV. The effects on neuromuscular function in
normal subjects and in myasthenia gravis. Bull Johns Hopkins Hosp 81:267–292.

Hassan RM, Pesce AJ, Sheng P, Hanenson IB (1981) Correlation of serum pseudocholinesterase and clinical course in two patients poisoned with organophosphate insecticides. Clin Toxicol 18:401–406.

Hawkins RD, Mendel B (1949) Studies on cholinesterase. VI. The selective inhibition of
true cholinesterase *in vivo*. Biochem J 44:260–264.

Hayes GR Jr, Funckes AJ, Hartwell WV (1964) Dermal exposure of human volunteers to
parathion. Arch Environ Health 8:829–833.

Hayes WJ (1983) Pesticides Studied in Man. Williams & Wilkins, Baltimore, pp 284–435.

Hayes WJ Jr (1969) Pesticides and human toxicity. Ann NY Acad Sci 160:40–54.

Heath DF, Park PO (1953) An irreversible choline-esterase inhibitor in white clover.
Nature (Lond) 172:206.

Hellenäs K-E, Nyman A, Slanina P, Lvvf L, Gabrielsson J (1992) Determination of potato
glycoalkaloids and their aglycone in blood serum by high-performance liquid chromatography. Application to pharmacokinetic studies in humans. J Chromatogr 573:69–78.

Hess AR, Angel RW, Barron KD, Bernsohn J (1963) Proteins and isozymes of esterases
and cholinesterases from sera of different species. Clin Chim Acta 8:656–667.

Hitchcock M, Murphy SD (1967) Enzymatic reduction of O,O-(4-nitrophenyl) phosphorothioate, O,O-diethyl O-(4-nitrophenyl) phosphate, and O-Ethyl O-(4-nitrophenyl)
benzene thiophosphonate by tissues from mammals, birds, and fishes. Biochem Pharmacol 16:1801–1811.

Hodgkin WE, Giblett ER, Levine H, Bauer W, Motulsky AG (1965) Complete pseudocholinesterase deficiency: genetic and immunologic characterization. J Clin Invest
44:486–493.

Hodgson E, Levi PE (1992) The role of the flavin-containing monooxygenase (EC
1.14.13.8) in the metabolism and mode of action of agricultural chemicals. Xenobiot-
ica 22:1175–1183.

Hodgson E, Levi PE (1994) Introduction to Biochemical Toxicology. Appleton & Lange,
Norwalk, CT.

Hodgson E, Rose RL, Ryu DY, Falls G, Blake BL, Levi PE (1995) Pesticide-metabolizing
enzymes. Toxicol Lett 82/83:73–81.

Hollingworth RM (1969) Dealkylation of organophosphorus esters by mouse liver
enzymes in vitro and in vivo. J Agric Food Chem 17:987–996.

Hollingworth RM (1970) The dealkylation of organophosphorus triesters by liver
enzymes. In: O'Brien RD, Yamamoto I (eds) Biochemical Toxicology of Insecticides,
Vol. 70. Academic Press, New York, pp 75–92.

Hollingworth RM (1971) Comparative metabolism and selectivity of organophosphate
and carbamate insecticides. Bull World Health Org 44:155–170.

Hollingworth RM, Metcalf RL, Fukuto TR (1967) The selectivity of sumithion compared
with methyl parathion. Metabolism in the white mouse. J Agric Food Chem 15:242–
249.

Hollingworth RM, Alstott RL, Litzenberg RD (1973) Glutathione *S*-aryl transferase in the metabolism of parathion and its analogs. Life Sci 13:191–199.

Holmstedt B (1963) Structure-activity relationships of the organophosphorus anticholinesterase agents. In: Eichler O, Farah A, Koelle GB (eds) Handbuch Der Experimentellen Pharmakologie. Springer, Heidelberg, pp 428–485.

Houwelingen RV, Nordoy A, van der Beek E, Houtsmuller U, de Metz M, Hornstra G (1987) Effect of a moderate fish intake on blood pressure, bleeding time, hematology, and clinical chemistry in healthy males. Am J Clin Nutr 46:424–436.

Hutson DH, Pickering BA, Donniger C (1968) Non-hydrolytic detoxification of insecticidal phosphate triester. In: Abstracts of the 5th Meeting of the Federation of European Biochemical Societies, Prague.

Johns RJ (1962) Familial reduction in red-cell cholinesterase. New Engl J Med 267:1344–1348.

Juul P (1968) Human plasma cholinesterase isoenzymes. Clin Chim Acta 19:205–213.

Kalow W, Davies RO (1958) The activity of various esterase inhibitors towards atypical human serum cholinesterase. Biochem Pharmacol 1:183–192

Kalow W, Genest K (1957) A method for the detection of atypical forms of human serum cholinesterase. Determination of dicubaine numbers. Can J Biochem Physiol 35:339–346.

Kalow W, Gunn DR (1959) Some statistical data on atypical cholinesterase of human serum. Ann Hum Genet 23:239–250.

Kalow W, Staron N (1957) On distribution and inheritance of atypical forms of human serum cholinesterase, as indicated by dibucaine numbers. Can J Biochem Physiol 35:1305–1321.

Karczmar AG (1970) Reactions of cholinesterases with substrate inhibitors and reactivators. In: International Encyclopedia of Pharmacology and Therapeutics: Anticholinesterase Agents. Pergamon Press, New York, pp 20–44.

Karczmar AG (1984) Acute and long lasting central actions of organophosphorus agents. Fundam Appl Toxicol 4:S1–S17.

Knaak JB, O'Brien RD (1960) Effect of EPN on in vivo metabolism of malathion by the rat and dog. J Agric Food Chem 8:198–203.

Knaak JB, Maddy KT, Gallo MA, Lillie DT, Craine EM, Serat WF (1978a) Worker reentry study involving phosalone application to citrus groves. Toxicol Appl Pharmacol 46:363–374.

Knaak JB, Maddy KT, Jackson T, Fredrickson AS, Peoples SA, Love R (1978b) Cholinesterase activity in blood samples collected from field workers and nonfield workers in California. Toxicol Appl Pharmacol 45:755–770.

Knaak JB, Peoples SA, Jackson TA, Fredrickson AS, Enos R, Maddy KT, Bailey JB, Düsch ME, Gunther FA, Winterlin WL (1978c) Reentry problems involving the use of dialifor on grapes in the San Joaquin Valley of California. Arch Environ Contam Toxicol 7:465–481.

Knaak JB, Iwata Y, Maddy KT (1989) The worker hazard posed by reentry into pesticide-treated foliage: development of safe reentry times, with emphasis on chlorthiophos and carbosulfan. In: Pautenbach DJ (ed) The Risk Assessment of Environmental Hazards: A Textbook Case of Studies. Wiley, New York.

Knaak JB, Al-Bayati MA, Raabe OG (1993a) Physiologically based pharmacokinetic modeling to predict tissue dose and cholinesterase inhibition in workers exposed to organophosphorous and carbamate pesticides. In: Wang GM, Knaak JB, Maibach HI (eds) Health Risk Assessment. Dermal and Inhalation Exposure and Absorption of Toxicants. CRC Press, Boca Raton, FL, pp 3–29.

Knaak JB, Al-Bayati MA, Raabe OG, Blancato JN (1993b) Development of *in vitro* V_{max} and K_m values for the metabolism of isofenphos by P-450 enzymes in animals and humans. Toxicol Appl Pharmacol 120:106–113.

Knaak JB, Al-Bayati MA, Raabe OG, Blancato JN (1994) Prediction of anticholinesterase activity and urinary metabolites of isofenphos. In: Saleh MA, Blancato JN, Nauman CH (eds) Biomarkers of Human Exposure to Pesticides. ACS Symposium Series. American Chemical Society, Washington, DC, pp 284–300.

Knaak JB, Al-Bayati MA, Raabe OG, Blancato JN (1996) Use of a multiple pathway and multiroute physiologically based pharmacokinetic model for predicting organophosphorus pesticide toxicity. In: Blancato JN, Brown RN, Dary CC, Saleh MA (eds) Biomarkers for Agrichemicals and Toxic Substances: Applications and Risk Assessment. ACS Symposium Series 643. American Chemical Society. Washington, DC, pp 206–228.

Koelle GB (1981) Organophosphate poisoning—an overview. Fundam Appl Toxicol 1:129–134.

Kozikowski AP, Xia Y, Reddy ER, Tickmantel W, Hanin I, Tang XC (1991) Synthesis of huperzine A and its analogues and their anticholinesterase activity. J Org Chem 56:4636–4645.

Krause A, Lane AB, Jenkins T (1988) A new high activity plasma cholinesterase variant. J Med Genet 25:677–681.

Krieger RI (1995) Pesticide exposure assessment. Toxicol Lett 82/83:65–72.

Kunstling TR, Rosse WF (1969) Erythrocyte acetylcholinesterase deficiency in paroxysmal nocturnal hemoglobinuria (PNH)—A comparison of the complement-sensitive and insensitive populations. Blood 33:607–616.

Kurono Y, Maki T, Yotsuyanagi Y, Ikeda K (1979) Esterase-like activity of human serum albumin: structure-activity relationships for the reactions with phenyl acetates and *p*-nitrophenyl esters. Chem Pharm Bull (Tokyo) 27:2781–2786.

Kurono Y, Kondo T, Ikeda K (1983) Esterase-like activity of human serum albumin: enantioselectivity in the burst phase of reaction with *p*-nitrophenyl α-methoxyphenyl acetate. Arch Biochem Biophys 227:339–341.

Kurono Y, Yamada H, Hata H, Okada Y, Takeuchi T, Ikeda K (1984) Esterase-like activity of human serum albumin. IV. Reactions with substituted aspirins and 5-nitrosalicyl esters. Chem Pharm Bull (Tokyo) 32:3715–3719.

Kurono Y, Miyajima M, Tsuji T, Yano T, Takeuchi T, Ikeda K (1991) Esterase-like activity of human serum albumin. VII. Reaction with *p*-nitrophenyl 4-guanidinobenzoate. Chem Pharm Bull (Tokyo) 39:1292–1294.

Kusic R, Jovanovic D, Randjelovic S, Joksovic D, Todorovic V, Boskovic B, Jokanovic M, Vojvodic V (1991) HI-6 in man: efficacy of the oxime in poisoning by organophosphorous insecticides. Hum Exp Toxicol 10:113–118.

La Du BN, Bartels CF, Nogueira CP, Hajra A, Lightstone H, Van Der Spek A, Lockridge O (1990) Phenotypic and molecular biological analysis of human butyrylcholinesterase variants. Clin Biochem 23:423–431.

La Du BN, Bartels CF, Nogueira CP, Arpagaus M, Lockridge O (1991) Proposed nomenclature for human butyrylcholinesterase genetic variants identified by DNA sequencing. Cell Mol Neurobiol 11:79–89.

LaMotta RV, Woronick CL (1971) Molecular heterogeneity of human serum cholinesterase. Clin Chem 17:135–144.

LaMotta RV, Williams HM, Wetstone HJ (1957) Studies of cholinesterase activity. II. Serum cholinesterase in hepatitis and cirrhosis. Gastroenterology 33:50–56.

LaMotta RV, McComb RB, Wetstone HJ (1965) Isozymes of serum cholinesterase: a new polymerization sequence. Can J Physiol Pharmacol 43:313–318.

LaMotta RV, McComb RB, Noll CR Jr, Wetstone HJ, Reinfrank RF (1968) Multiple forms of serum cholinesterase. Arch Biochem Biophys 124:299–305.

Lapidot-Lifson Y, Prody CA, Ginzberg D, Meytes D, Zakut H, Soreq H (1989) Coamplification of human acetylcholinesterase and butyrylcholinesterase genes in blood cells: correlation with various leukemias and abnormal megakaryocytopoiesis. Proc Natl Acad Sci USA 86:4715–4719.

Layer PG, Willbold E (1995) Novel functions of cholinesterases in development, physiology and disease. Prog Histochem Cytochem, 29:1–94 reverse Fischer, Stuttgart.

Li Y, Camp S, Rachinsky TL, Getman D, Taylor P (1991) Gene structure of mammalian acetylcholinesterase. J Biol Chem 266:23083–23090.

Li Y, Camp S, Taylor P (1993a) Tissue-specific expression and alternative mRNA processing of the mammalian acetylcholinesterase gene. J Biol Chem 268:5790–5797.

Li Y, Camp S, Rachinsky TL, Bongiorno C, Taylor P (1993b) Promoter elements and transcriptional control of the mouse acetylcholinesterase gene. J Biol Chem 268:3563–3572.

Linquist DA, Bull DL (1967) Fate of 3-hydroxy-*n*-methyl-*cis*-crotonamide dimethyl phosphate in cotton plants. J Agric Food Chem 15:267–272.

Lockridge O (1990) Genetic variants of human serum cholinesterase influence metabolism of the muscle relaxant succinylcholine. Pharm Exp Ther 47:35–60.

Lockridge O, La Du BN (1978) Comparison of atypical and usual human serum cholinesterase. J Biol Chem 253:361–366.

Lockridge O, Eckerson HW, LaDu BN (1979) Interchain disulfide bonds and subunit organization in human serum cholinesterase. J Biol Chem 254:8324–8330.

Lockridge O, Adkins S, LaDu BN (1987a) Location of disulfide bonds within the sequence of human serum cholinesterase. J Biol Chem 262:12945–12952.

Lockridge O, Bartels CF, Vaughan TA, Wong CK, Norton SE, Johnson LL (1987b) Complete amino acid sequence of human serum cholinesterase. J Biol Chem 262:549–557.

Loft AGR (1990) Determination of amniotic fluid acetylcholinesterase activity in the antenatal diagnosis of foetal malformations: the first ten years. J Clin Chem Clin Biochem 28:893–911.

Loft AGR, Mortensen V, Hangaard J, Norgaard-Pedersen B (1991) Ratio of immunochemically determined amniotic fluid acetylcholinesterase to butyrylcholinesterase in the differential diagnosis of fetal abnormalities. Br J Obstet Gynaecol 98:52–56.

Long KR (1975) Cholinesterase activity as a biological indicator of exposure to pesticides. Int Arch Occup Environ Health 36:75–86.

Lucier GW, Menzer RE (1968) Metabolism of dimethoate in bean plants in relation to its mode of application. J Agric Food Chem 16:936–945.

Lucier GW, Menzer RE (1970) Nature of oxidative metabolites of dimethoate formed in rats, liver microsomes and bean plants. J Agric Food Chem 18:698–704.

Ludwig PD, Kilian DJ, Dishburger HJ, Edwards HN (1970) Results of human exposure to thermal aerosols containing dursban insecticide. Mosq News 30:346–354.

Lynch TJ, Maltes CE, Singh A, Bradley RM, Brady RO, Dretchen KL (1997) Cocaine detoxification by human plasma by tyrylcholinesterase. Toxicol Appl Pharmacol 145:363–371.

Ma T, Chambers JE (1994) Kinetic parameters of desulfuration and dearylation of parathion and chlorpyrifos by rat liver microsomes. Food Chem Toxicol 32:763–767.

Main AR (1960a) The purification of the enzyme hydrolyzing diethyl p-nitrophenyl phosphate (paraoxon) in sheep serum. Biochem J 74:10–20.

Main AR (1960b) The differentiation of the A-type esterases in sheep serum. Biochem J 75:188–195.

Main AR (1964) Affinity and phosphorylation constants for the inhibition of esterases by organophosphates. Science 144:992–993.

Main AR (1969) Kinetic evidence of multiple reversible cholinesterases based on inhibition by organophosphates. J Biol Chem 244:829–840.

Main AR (1973) Kinetics of active-site directed irreversible inhibition. In: Hayes WL (ed) Essays in Toxicology. Academic Press, New York, p 59.

Main AR, Iverson F (1966) Measurement of the affinity and phosphorylation constants governing irreversible inhibition of cholinesterases by di-isopropyl phosphorofluoridate, Biochem J 100:525–531.

Masson P (1979) Formes moleculaires multiples de la butyrylcholinesterase du plasma humain. I. Parametres moleculaires apparents et ebauche de la structure quaternaire. Biochim Biophys Acta 578:493–504.

Masson P (1989) A naturally occurring molecular form of human plasma cholinesterase is an albumin conjugate. Biochim Biophys Acta 988:258–266.

Masson P (1991) Molecular heterogeneity of human plasma cholinesterase. In: Massoulii J, Bacou F, Barnard E, Chatonnet A, Doctor B Quinn DM (eds) Cholinesterases: Structure, Function, Mechanism, Genetics, and Cell Biology. American Chemical Society, Washington DC, pp 42–46.

Matsumura R (1975) Toxicology of Insecticides, Plenum Press, New York.

Matsumura F, Hogendijk CJ (1964) The enzymatic degradation of parathion in organophosphate-susceptible and -resistant houseflies. J Agric Food Chem 12:447–452.

Matsuzaki S, Iwamura K, Katsunuma T, Kamiguchi H (1980a) Separation of serum cholinesterase isozymes by an improved polyacrylamide gel electrophoresis and its application for the study of liver diseases (Part I). Gastroenterol Jpn 15:33–40.

Matsuzaki S, Iwamura K, Katsunuma T, Kamiguchi H (1980b) Abnormalities of serum cholinesterase isozyme in liver cirrhosis and hepatoma (Part II). Gastroenterol Jpn 15:543–549.

Mattes CE, Lynch TJ, Singh A, Bradley RM, Kellaris PA, Brady PO, Dretchen KL (1997) Therapeutic use of butyrylcholinesterase for cocaine intoxication. Toxicol Appl Pharmacol 145:372–380.

McCurdy SA, Hansen ME, Weisskopf CP, Lopez RL, Schneider F, Spencer J, Sanborn JR, Krieger RI, Wilson BW, Goldsmith DF, Schenker MB (1994) Assessment of azinphosmethyl exposure in California peach harvest workers. Arch Environ Health 49:289–296.

Menzer R, Casida JE (1965) Nature of toxic metabolites formed in mammals insects and plants from 3-(dimethoxyphosphinyloxy)-N, N-dimethyl-cis-crotonamide and its N-methyl analog. J Agric Food Chem 13:102–112.

Menzie CM (1966) Metabolism of pesticides. Special Scientific Report—Wildlife No. 96. U. S. Department of the Interior, Fish and Wildlife Service, Washington, DC.

Menzie CM (1969) Metabolism of pesticides. Special Scientific Report—Wildlife No. 127. U.S. Bureau of Sport Fisheries and Wildlife, Washington, DC.

Mertens HW, Lewis MF, Steen JA (1974) Some behavioral effects of pesticides: phosdrin and free-operant escape-avoidance behavior in gerbils. Aerospace Med 45:1171–1176.

Metcalf RL, March RB (1953) The isomerization of organic thiophosphate insecticides. J Econ Entomol 46:288–294.

Metcalf RL, Fukuto TR, March RB (1957) Plant metabolism of dithio-systox and thimet. J Econ Entomol 50:338–345.

Michel HO, Krop S (1951) The reaction of cholinesterase with diisopropyl fluorophosphate. J Biol Chem 190:119–125.

Mileson BE, Chambers JE, Chen WL, Dettbarn W, Ehrich M, Eldefrawi AT, Gaylor DW, Hamernik K, Hodgson EA, Karczmar G, Padilla S, Pope, CN, Richardson RJ, Saunders DR, Sheets LP, Sultatos LG, Wallace KB (1998) Common mechanism of toxicity: a case study of organophosphorous pesticides. Toxicol Sci 41:8–20.

Moeller HC, Rider JA (1962) Plasma and red blood cell cholinesterase activity as indications of the threshold of incipient toxicity of ethyl-p-nitrophenyl thionobenzenephosphonate (EPN) and malathion in human beings. Toxicol Appl Pharmacol 4:123–130.

Moore DH, Clifford CB, Crawford IT, Cole GM, Baggett JM (1995) Review of nerve agent inhibitors and reactivators of acetylcholinesterase. In: Quinn _, et al. (eds) Enzymes of the Cholinesterase Family. Plenum Press, New York.

Morgan DP (ed) (1982) Recognition and Management of Pesticide Poisonings, 3rd Ed. U.S. Environmental Protection Agency, Washington, DC pp 1–8.

Moriearty PL, Becker RE (1993) Inhibition of human brain and RBC acetylcholinesterase (AChE) by heptylphysostigmine (HPTL). Methods Find Exp Clin Pathol 14:615–621.

Morris SC, Lee TH (1984) The toxicity and teratogenicity of Solanaceae glycoalkaloids, particularly those of the potato (Solanum tuberosum): a review. Food Technol Aust 36:118–124.

Moss DE, Kobayashi H, Pacheco G, Palacios R, Perez RDG (1988) Methanesulfonyl fluoride: A CNS selective cholinesterase inhibitor. In: Giacobini E, Becker R (eds) Current Research in Alzheimer Therapy. Taylor & Francis, New York, pp 305–314.

Mounter LA (1956) Identity of diisopropylfluorophosphatase and acylase. Fed Proc 15:317–318.

Mücke W, Alt KO, Esser O (1970) Degradation of ^{14}C-labelled diazinon in the rat. J Agric Food Chem 18:208–212.

Murphy SD (1966) Liver metabolism and toxicity of thiophosphate insecticides in mammalian, avian and piscine species. Proc Soc Exp Biol Med 123:392–398.

Murphy SD (1972) The toxicity of pesticides and their metabolites. In: Degradation of Synthetic Organic Molecules in the Biosphere. National Academy of Science, Washington, DC, pp 313–335.

Murphy SD (1980) Pesticides. In: Casarett LJ, Doull J (eds) Casarett and Doull's Toxicology: The Basic Science of Poisons, 2nd Ed. Macmillan, New York, pp 357–375.

Murphy SD, Lauwerys RR, Cheever KL (1968) Comparative anticholinesterase action of organophosphorus insecticides in vertebrates. Toxicol Appl Pharmacol 12:22–35.

Nakatsugawa T, Dahm PA (1967) Microsomal metabolism of parathion. Biochem Pharmacol 16:25–38.

Namba T (1971) Cholinesterase inhibition by organophosphorus compounds and its clinical effects. Bull Org Mond Sante (Bull World Health Org) 44:289–307.

Namba T, Hiraki K (1958) PAM (pyridine-2-aldoxime methiodide) therapy for alkylphosphate poisoning. JAMA 166:1834–1839.

Namba T, Nolte CT, Jackrel J, Grob D (1971) Poisoning due to organophosphate insecticides. Am J Med 50:475–492.

Neal RA (1967a) Studies on the metabolism of diethyl 4-nitrophenyl phosphorothionate (Parathion) *in vitro*. Biochem J 103:183–191.

Neal RA (1967b) Studies of the enzymic mechanism of the metabolism of diethyl 4-nitrophenyl nitrophenyl phosphorothionate (parathion) by rat liver microsomes. Biochem J 105:289–297.

Neitlich HW (1966) Increased plasma cholinesterase activity and succinylcholine resistance: a genetic variant. J Clin Invest 45:380–387.

Neville LF, Gnatt A, Loewenstein Y, Soreq H (1990) Aspartate-70 to glycine substitution confers resistance to naturally occurring and synthetic anionic-site ligands on in-ovo produced human butyrylcholinesterase. J Neurosci Res 27:452–460.

Nigg HN, Olexa M (1986) Safety and health maintenance. In: IFAS Pesticides Policies and Procedures. University of Florida, Gainesville.

Nigg HN, Ramos LE, Graham ME, Sterling J, Brown S, Cornell JA (1996) Inhibition of human plasma and serum butyrylcholinesterase (EC 3.1.1.8) by α-chaconine and α-solanine. Fundam Appl Toxicol 33:272–281.

NIOSH (1990) Pocket Guide to Chemical Hazards. DHHS (NIOSH) publication no. 90–117. Publications Dissemination, NIOSH, Cincinnati, OH.

O'Brien RD (1957) Properties and metabolism in the cockroach and mouse of malathion and malaoxon. J Econ Entomol 50:159–164.

O'Brien RD (1960) Toxic Phosphorus Esters: Chemistry, Metabolism, and Biological Effects. Academic Press, New York.

O'Brien RD, Kimmel EC, Sferra RP (1965) Toxicity and metabolism of famphur in insects and mice. J Agric Food Chem 13:366–372.

O'Malley MA, McCurdy SA (1990) Subacute poisoning with phosalone, an organophosphate insecticide. West J Med 153:619–624.

Oosterbaan RA, Kunst P, Cohen JA (1955) Nature of the reaction between diisopropyl fluorophosphate and chymotrypsin. Biochem Biophys Acta 16:299–300.

Orgell WH (1963) Inhibition of human plasma cholinesterase *in vitro* by alkaloids, glycosides, and other natural substances. Lloydia (Cinci) 26:36–43.

Orgell WH, Hibbs ET (1963) Cholinesterase inhibition *in vitro* by potato foliage extracts. Am Potato J 40:403–405.

Orgell WH, Vaidya KA, Dahm PA (1958) Cholinesterase inhibition *in vitro* by extracts of potato. Iowa Acad Sci 65:160–162.

Orgell WH, Vaidya KA Hamilton EW (1959) A preliminary survey of some midwestern plants for substances inhibiting human plasma cholinesterase *in vitro*. Proc Iowa Acad Sci 66:149–154.

Padilla S, Lassiter L, Crofton K, Moser VC (1996) Blood cholinesterase activity: Inhibition as an indicator of adverse effect. In: Blancato JN, Brown RN, Dary CC, Saleh MA (eds) Biomarkers for Agrochemicals and Toxic Substances: Applications and Risk Assessment. ACS Symposium Series 643. American Chemical Society, Washington, DC, pp 70–78.

Pardue JR, Hansen EA, Barron RP, Chen JYT (1970) Diazinon residues on field-sprayed kale. Hydroxydiazinon—a new alteration product of diazinon. J Agric Food Chem 18:405–408.

Parke DV (1968) Radioisotopes in the study of the metabolism of foreign compounds. In: Roth LJ (ed) Isotopes in Experimental Pharmacology 1965:315–342.

Pasquet J., Mazuret A, Fournel J, Koenig FH (1976) Acute oral and percutaneous toxicity of phosalone in the rat, in comparison with azinphosmethyl and parathion. Toxicol Appl Pharmacol 37:85–92.

Patil BC, Sharma RP, Salunkhe DK, Salunkhe K (1972) Evaluation of solanine toxicity. Food Cosmet Toxicol 10:395–398.

Peakall D (1992) Animal Biomarkers as Pollution Indicators. Chapman & Hall, London, pp 1–290.

Pesticide & Toxic Chemical News (1991) Cholinesterase assay recommendations may come by year end. Pesticide & Toxic Chemical News, December 11, 1991, pp 29–32.

Polhuijs M, Langenberg JP, Benschop HP (1997) New method for retrospective detection of exposure to organophosphorous anticholinesterases: application to alleged sarin victims of Japanese terrorists. Toxicol Appl Pharmacol 146:156–161.

Prester L, Simeon V (1991) Kinetics of the inhibition of human serum cholinesterase phenotypes with the dimethylcarbamate of (2-hydroxy-5-phenylbenzyl)-trimethylammonium bromide (Ro 02-0683). Biochem Pharmacol 42:2313–2316.

Raabe OG, Knaak JB, Al-Bayati MA, Enslein K, Gombar VK (1994) Mechanistically-based alternative methods in toxicology structure-activity and PBPK/PBPD models in toxicology, Research proposal, University of California at Davis, Davis, CA.

Radic Z, Quinn RD, Vellom DC, Camp S, Taylor P (1995) Amino acid residues in acetylcholinesterase which influence fasciculin inhibition. In: Quinn DM, et al (eds) Enzymes of the Cholinesterase Family. Plenum Press, New York, pp 183–188.

Raffaele KC, Rees C (1990) Neurotoxicology dose/response assessment for several cholinesterase inhibitors: use of uncertainty factors. Neurotoxicology 11:237–256.

Reiner E, Aldridge WN, Hoskins FCG (1989) Enzymes Hydrolysing Organophosphorus Compounds. Ellis Horwood, Chichester, England.

Reiter LW, Talens GM, Woolley DE (1975) Parathion administration in the monkey: time course of inhibition and recovery of blood cholinesterases and visual discrimination performance. Toxicol Appl Pharmacol 33:1–13.

Rider JA, Hodges JL Jr, Swader J, Wiggins AD (1957) Plasma and red cell cholinesterase in 800 "healthy" blood donors. J Lab Clin Med 50:376–383.

Rider JA, Moeller HC, Puletti EJ, Swader JI (1969) Toxicity of parathion, systox, octamethyl pyrophosphoramide, and methyl parathion in man. Toxicol Appl Pharmacol 14:603–611.

Rubinstein HM, Dietz AA, Hodges LK, Lubrano T, Czebotar V (1970) Silent cholinesterase gene: variations in the properties of serum enzyme in apparent homozygotes. J Clin Invest 49:479–486.

Sanderson DM, Edson EF (1964) Toxicological properties of the organophosphorous insecticide dimethoate. Br J Ind Med 21:52–64.

Schaffer NK, Michel HO, Bridges AF (1973) Amino acid sequence in the region of the reactive serine residue of eel acetylcholinesterase. Biochemistry 12:2946–2950.

Schumacher M., Maulet Y, Camp S, Taylor P (1988) Multiple messenger RNA species give rise to the structural diversity in acetylcholinesterase. J Biol Chem 263:18979–18987.

Schüürman G (1992) Ecotoxicology and structure-activity studies of organophosphorus compounds. In: Rational Approaches to Structure, Activity, and Ecotoxicity of Agrochemicals, eds. Draber W and Toshia F, CRC Press, Boca Raton, Fl, 1992.

Sheets LP, Hamilton BF, Sangha GK, Thyssen JH (1997) Subchronic neurotoxicity screening studies with six organophosphate insecticides: an assessment of behavior and morphology related to cholinesterase inhibition. Fundam Appl Toxicol 35:101–119.

Sherman KA, Kumar V, Ashford JW, Murphy JW, Elble RJ, Giacobini E (1988) Effect of oral physostigmine in senile dementia patients: utility of blood cholinesterase inhibition and neuroendocrine responses to define pharmacokinetics and pharmacodynamics. In: Strong R, Wood WG (eds) Central Nervous System Disorders of Aging: Clinical Intervention and Research. Raven Press, New York, pp 71–90.

Sinden SL, Webb RE (1972) Effect of variety and location on the glycoalkaloid content of potatoes. Am Potato J 49:334–338.

Smith RL, Williams RT (1966) Implications of the conjugation of drugs and other exogenous compounds. In: Dutton GJ (ed) Glucuronic Acid, Free and Combined. Academic Press, New York, pp 457–491.

Soreq H, Zakut H (1993) Human Cholinesterases and Anticholinesterases. Academic Press, New York.

Spencer EY, O'Brien RD, White RW (1957) Permanganate oxidation products of schradan. J Agric Food Chem 5:123–127.

Srinivasan R, Karczmar AG, Bernshon J (1976) Rat brain acetylcholinesterase and its isoenzymes after intracerebral administration of DFP. Biochem Pharmacol 25:2739–2745.

Stefanini M (1985) Enzymes, isozymes, and enzyme variants in the diagnosis of cancer. Cancer (Phila) 55:1931–1936.

Strelitz F (1944) Studies on cholinesterase. IV. Purification of pseudocholinesterase from horse serum. Biochem J 38:86–88.

Sultatos LG (1990) A physiologically based pharmacokinetic model of parathion based on chemical-specific parameter determined in vitro. J Am Coll Toxicol 9:611–619.

Sultatos LG, Gagliardi CL (1990) Desulfuration of the insecticide parathion by human placenta in vitro. Biochem Pharmacol 39:799–801.

Sultatos LG, Murphy SD (1983) Kinetic analyses of the microsomal biotransformation of the phosphorothioate insecticides chlorpyrifos and parathion. Fundam Appl Toxicol 3:16–21.

Sumerford WT, Hayes WJ Jr, Johnston JM, Walker K, Spillane J (1953) Cholinesterase response and symptomatology from exposure to organic phosphorus insecticides. Arch Ind Hyg Occup Med 7:383–398.

Sussman JL, Harel M, Frolow F, Oefner C, Goldman A, Toker L, Silman L (1991) Atomic structure of acetylcholinesterase from *Torpedo californcia*: a prototypic acetylcholine-binding protein. Science 253:872–879.

Sutton LD, Froelich S, Hendrickson HS, Quinn DM (1991) Cholesterol esterase catalyzed hydrolysis of mixed micellar thiophosphatidylcholines: a possible charge-relay mechanism. Biochemistry 30:5888–5893.

Svensmark O (1963) Enzymatic and molecular properties of cholinesterases in human liver. Acta Physiol Scand 59:378–389.

Tammelin LE (1958a) Choline esters. Substrates and inhibitors of cholinesterase. Svensk Kem Tidskr 70:157–181.

Tammelin LE (1958b) Dialkoxyphosphorylthiocholines, alkoxymethylphosphorylthiocholines and analogous choline esters. Synthesis, pK_a of tertiary homologs and cholinesterase inhibition. Acta Chem Scand 11:1340–1349.

Tammelin LE (1958c) Organophosphorylcholines and cholinesterases. Arkiv Kemi 12:287–298.

Tang XC, Zhu XD, Lu WH (1988) Studies on the nootropic effects of huperzine A and B: two selective AChE inhibitors. In: Giacobina E, Becker R (eds) Current Research in Alzheimer Therapy. Taylor & Francis, New York, pp 289–293.

Thomsen T, Kewitz H (1990) Selective inhibition of human acetylcholinesterase by galanthamine *in vitro* and *in vivo*. Life Sci 46:1553–1558.

Tietz NW, Finley PR, Pruden EL (eds) (1990) Clinical Guide to Laboratory Tests, 2nd Ed. Saunders, Philadelphia.

U.S. Environmental Protection Agency (1991) Pesticide assessment guideline, subdivision F. Hazard evaluation: human and domestic animals. Addendum 10: Neurotoxicity, series 81, 82, and 83. USEPA 540/09-91-123.Office of Prevention, Pesticides and Toxic Substances. Washington, DC.

U.S. Environmental Protection Agency (1992) Worker protection standards. USEPA, Washington, DC, p. 36.

Uchida T, O'Brien RD (1967) Dimethoate degradation by human liver and its significance for acute toxicity. Toxicol Appl Pharmacol 10:89–94.

Uchida T, Dauterman WS, O'Brien RD (1964) The metabolism of dimethoate by vertebrate tissues. J Agric Food Chem 12:48–52.

Ursdin E (1970) Reactions of cholinesterases with substrate inhibitors and reactivators. In: International Encyclopedia of Pharmacology and Therapeutics: Anticholinesterase Agents, Pergamon Press, New York, pp 49–354.

Vandekar M (1980) Minimizing occupational exposure to pesticides: cholinesterase determination and organophosphorus poisoning. Residue Rev 75:67–80.

Van Lith HA, Herman S, Zhang X, Van Der Palen JGP, Van Zutphen LFM, Beynen AC (1990) Influence of dietary fats on butyrylcholinesterase and esterase-1 (ES-1) activity in plasma of rats. Lipids 25:779–786.

Vincent D, Parant M (1956) Atropine, apo-atropine et cholinestirases. Comptes Rendus Seances Soc Biol Toulouse 150:444–447.

Walker CH (1989) The development of an improved system of nomenclature and classification of esterases. In: Reiner E, Aldridge WN, Hoskin FCG (eds) Enzymes Hydrolyzing Organophosphorus Compounds. Halsted Press, New York, 266 pp.

Wallace KB, Dargan JE (1987) Interinsic metabolic clearance of parathion and paraoxon by livers from fish and rodeasts. Toxicol Appl Pharmacol 90:235–242.

Wallace KD (1992) Species-selective toxicity of organophosphorus insecticides: a pharmacodynamic phenomenon. In: Chambers JE, Levi PE (eds) Organophosphates, Chemistry, Fate, and Effects. Academic Press, San Diego, pp 79–105.

Wang C, Murphy SD (1982) Kinetic analysis of species difference in acetylcholinesterase sensitivity to organophosphate insecticides. Toxicol Appl Pharmacol 66:409–419.

Wester RC, Maibach HI, Melendres J, Sedik L, Knaak JB, Wang R (1992) In vivo and in vitro percutaneous absorption and skin evaporation of isofenphos in man. Fundam Appl Toxicol 19:521–526.

Wetstone HJ, LaMotta RV (1965) The clinical stability of serum cholinesterase activity. Clin Chem 11:653–663.

Whetstone RR, Phillips DD, Sun YP, Ward LF, Shellenberger TE (1966) 2-Chloro-1-(2,4,5-trichlorophenyl) vinyl dimethyl phosphate, a new insecticide with low toxicity to mammals. J Agric Food Chem 14:352–356.

Whittaker M (1986) Cholinesterase. Monographs in Human Genetics, Vol. II. Karger, Basel, Switzerland.

Wilkinson GN (1961) Statistical estimations in enzyme kinetics. Biochem J 80:324–332.

Wilson BW, Hooper M, Chow E, Higgins R, Knaak JB (1984) Antidotes and neuropathic potential of isofenphos. Bull Environ Contam Toxicol 33:386–394.

Wilson BW, Sanborn JR, O'Malley MA, Henderson JD, Billitti JR (1997) Monitoring the pesticide-exposed worker. Occup Med State of the Art Reviews, 12:347–363.

Wilson BW, Billitti JE, Henderson JD, McCarthy SA, O'Malley MA, Sanborn JR, McCurdy SA (1998) Optimizing cholinesterase assays for monitoring humans. Bull Environ Contam Toxicol (in press).

Wilson GS (1959) A small outbreak of solanine poisoning. Monthly Bull Med Res Counc 18:207–210.

Wilson IB (1951) Acetylcholinesterase. XI. Reversibility of tetraethyl pyrophosphate inhibition. J Biol Chem 190:111–117.

Wilson IB (1960) Acetylcholinesterase. In: Boyer PD, Lardy H, Myrbdck K (eds) The Enzymes, 2nd Ed. Academic Press, New York, pp 501–520.

Wilson IB, Bergmann F (1950) Acetylcholinesterase. VIII. Dissociation constants of the active groups. J Biol Chem 186:683–692.

Wilson IB, Bergmann F, Nachmansohn D (1950) Acetylcholinesterase. X. Mechanism of the catalysis of acylation reactions. J Biol Chem 188:781–790.

Witter RF (1963) Measurement of blood cholinesterase. Arch Environ Health 6:537–563.

Wolthius OL, Vanwersch RAP (1984) Behavioral changes in the rat after low doses of cholinesterase inhibitors. Fundam Appl Toxicol 4:S195–S208.

Wustner DA, Fukuto TR (1974) Affinity and phosphorylation constants for the inhibition of cholinesterase by the optical isomers of O-2-butyl S-2-(dimethylammonium) ethyl ethylphosphorothioate hydrogen oxalate. Pestic Biochem Physiol 4:365–376.

Yang RSH, Dauterman WC, Hodgson E (1971a) Metabolism *in vitro* of diazinon and diazoxon in rat liver. J Agric Food Chem 19:10–13.

Yang RSH, Dauterman WC, Hodgson E (1971b) Metabolism *in vitro* of diazinon and diazoxon in susceptible and resistant houseflies. J Agric Food Chem 19:14–19.

Yoshida A, Motulsky AG (1969) A pseudocholinesterase variant (E Cynthiana) associated with elevated plasma enzyme activity. Am J Hum Genet 21:486–498.

Yoshida K, Kurono Y, Mori Y, Ikeda K (1985) Esterase-like activity of human serum albumin. V. Reaction with 2,4-dinitrophenyl diethyl phosphate. Chem Pharm Bull (Tokyo) 33:4995–5001.

Yu Q-S, Liu C, Brzostowska M, Chrisey L, Brossi A, Greig NH, Atack JR, Soncrant TT, Rapoport SI, Radunz H-E (1991) Physovenines: efficient synthesis of (-)- and (+)-physovenine and synthesis of carbamate analogues of (-)-physovenine. Anticholinesterase activity and analgesic properties of optically active physovenines. Helv Chim Acta 74:761–766.

Zakut H, Lieman-Hurwitz J, Zamir R, Sindell L, Ginzberg D, Soreq H (1991) Chorionic villus cDNA library displays expression of butyrylcholinesterase: putative genetic disposition for ecological danger. Prenatal Diagn 11:597–607.

Zavon MR (1965) Blood cholinesterase levels in organic phosphate intoxication. JAMA 192:137.

Zeller von EA, Bissegger A (1943) Uber die Cholin-esterase des gehirns und der Erythrocyten. Zugleich 3. Mitteilung uber die beeinflussung von fermentreaktionen durch. Chemotherapeutica und pharmaka. Helv Chim Acta 26:1619–1630.

Zhang HX, Sultatos LG (1991) Biotransformation of the organophosphorus insecticides parathion and methyl parathion in male and female rat livers perfused in situ. Drug Metab Dispos 19:473–477.

Zimmerman JK, Grothusen JR, Bryson PK (1989) Partial purification and characterization of paraoxonase from rabbit serum. In: Reiner E, Aldridge WN, Hoskin FCG (eds) Enzymes Hydrolysing Organophosphorus Compounds. Ellis Horwood, Chichester, England, pp 128–142.

Manuscript received April 15, 1998, accepted April 27 1999.

Rev Environ Contam Toxicol 163:113–186   © Springer-Verlag 2000

Toxicology and Risk Assessment of Freshwater Cyanobacterial (Blue-Green Algal) Toxins in Water

Tai Nguyen Duy, Paul K.S. Lam, Glen R. Shaw, and Des W. Connell

Contents

T. N. Duy
Faculty of Environmental Sciences, Griffith University, Nathan, Queensland, Australia.
P. K.S. Lam (⊠)
Department of Biology and Chemistry, City University of Hong Kong, Hong Kong.
G. R. Shaw
National Research Centre for Environmental Toxicology, Coopers Plains, Queensland 4108, Australia
D. W. Connell
School of Public Health, Griffith University, Logan, Queensland, Australia.

I. Introduction

The adverse effects of cyanobacterial toxins were first reported as stock deaths at Lake Alexandrina, South Australia, in 1878. Since then, cyanobacterial poisonings in animals and humans have been widely reported around the world (Codd and Poon 1988). In fact, cattle and wildlife mortality from cyanobacterial poisonings is relatively common in many countries (Carmichael 1981). Animals that have been killed in large numbers include cattle, sheep, pigs, birds, and fish; small numbers of deaths of horses, dogs, rodents, amphibians, and invertebrates have also been recorded (Codd and Poon 1988). According to the compilations of Carmichael (1992a), approximately 85 animal poisoning incidents related to cyanobacterial blooms have been recorded around the world from 1878 to 1991.

Human illnesses associated with exposure to cyanobacterial toxins have been widely reported in Australia (QWQTF 1992), Europe, Africa, Asia, and North America (Codd et al. 1989) and South America (Jochimsen et al. 1998). One of the most serious human poisonings caused by cyanobacterial toxins was the death from liver failure of more than 50 patients receiving contaminated dialysis water in Caruaru, Brazil, in 1996 (Jochimsen et al. 1998). Another human illness incident that has been attributed to cyanobacterial toxins occurred at Sewickley, PA. In this incident, 5000 people of a population of 8000 suffered diarrhea, abdominal cramps, and a variety of other symptoms after consuming water contaminated by toxic cyanobacteria (Falconer 1994a). Likewise, 139 children and 10 adults on Palm Island (Queensland, Australia) were believed to be affected by the presence of these toxins in water supplies in 1979. Nearly 70% of patients required intravenous therapy, and 40% developed profuse bloody diarrhea after 1 or 2 d exposure (Falconer 1993b). Other medical conditions attributed to the presence of cyanobacterial toxins in drinking water were recorded in East Africa (Falconer 1993b); Malpas Dam, New South Wales (Falconer et al. 1983); Peel-Harvey Estuary, Western Australia; and Charleston, WVA, and Monroe County, PA (Ressom et al. 1994).

Currently, there is a general lack of adequate risk assessment information for cyanobacterial toxins. The model of risk assessment for these toxins particularly is incomplete, mainly focusing on the relatively common toxin, microcystin, and ignoring other toxins. A few countries (e.g., Australia, Canada, Great Britain, and the United States) have provisional guidelines for evaluating the level of cyanobacterial toxins in drinking water. Because of possible adverse effects of cyanobacterial toxins on humans and animals, there is a need for a better understanding of the hazard posed by these toxins so as to assess the risks associated with them. This knowledge will provide a base to undertake risk communication and risk management for the toxins and will also contribute to establishing official guideline values of cyanobacterial toxins in water that will not pose an unacceptable risk to livestock and humans. Furthermore, it may contribute to improving the safety of drinking water supplies, particularly those in developing countries where a high proportion of the population may be consuming untreated water from lakes and rivers.

The overall aim of this review is to identify, assess, and predict risks to humans and animals (both wild and domestic) posed by freshwater cyanobacterial toxins. To achieve this, the review focuses on the following issues:

1. Review of the current knowledge related to freshwater cyanobacteria and their toxins
2. Assessment of the risks associated with freshwater cyanobacterial toxins
3. Establishment of safety guideline values of selected cyanobacterial toxins (anatoxin-a, microcystins, microcystin-LR, and cylindrospermopsin) based on available data in the scientific literature.

II. Characteristics of Cyanobacteria

A. Taxonomy of Cyanobacteria

Cyanobacteria (blue-green algae) belong to the kingdom Monera (Prokaryota), division Eubacteria, class Cyanobacteria (Ressom et al. 1994). The class includes about 150 genera and about 2000 species (Skulberg et al. 1993). Cyanobacteria were earlier classified as algae according to the Botanical Code, and this classification had been used for some 180 years (Carr and Whitton 1982). However, the advent of electron microscopy had indicated that cyanobacteria are distinct from other algae by their cytology, distribution, ecology, reproduction, physiology, and biochemistry. Because cyanobacteria possess some algal characteristics as well as some bacterial properties, they are classified in the algal class according to the principles of the International Code of Botanical Nomenclature (1972) while they also fit the bacterial nomenclature (Skulberg et al. 1993). The cyanobacteria are now placed within the group Eubacteria in the phylogenetic taxonomy, separate and apart from the Archaebacteria and eukaryotes, and two nomenclature systems are applied in parallel to classify Cyanobacteria (Skulberg et al. 1993).

B. Biology of Cyanobacteria

Cyanobacteria are gram-negative bacteria capable of producing a wide range of potent toxins as secondary metabolites (Codd 1994). They can be found in a wide range of habitats, from ice fields to hot springs and deserts. This group is one of the most morphologically, physiologically, and metabolically diverse groups of prokaryotes (Codd 1994). In terms of morphology, cyanobacteria are either single celled, colonial, or filamentous. Individual cells generally vary from 3 to 10 µm in diameter. Cyanobacterial cells can be round, oval, or globular and usually exist in solitary, free-living forms or platelike colonies of cells enclosed in mucilage and consisting of a few to thousands of cells (Van den Hoek et al. 1995). Some colonies of certain species can be seen with the naked eye (Baker 1991).

    Cyanobacteria are prokaryotes without nuclei. The absence of a nuclear membrane in their structure distinguishes them from other algae. In most species, each cell is surrounded by a cell wall made up by peptidoglycan and lipopolysaccharide layers that are in turn surrounded by a gelatinous or mucilaginous sheath (Ressom et al. 1994). The presence of a gelatinous sheath may enable cyanobacteria to survive in desiccation or other extreme conditions (Van den Hoek et al. 1995). The cytoplasm of cyanobacteria is heterogeneous and usually incorporates granular structures (Skulberg et al. 1993). These structures include glycogen granules, lipid globules, cyanophycin granules, and polyphosphate bodies (Van den Hoek et al. 1995). Some cyanobacteria possess gas vacuoles that provide these organisms with a buoyancy regulation mechanism (Ressom et al. 1994).

    Cyanobacteria contain chlorophyll *a*, carotene, xanthophyll, blue *c* phycocyanin, and red *c* phycoerythrin. The latter two pigments can only be found in cyanobacteria (Benson 1969). As the result of the presence of a diverse range of pigments, cyanobacterial color can be either olive green, grey-green, yellow-brown, or purplish to red (Ressom et al. 1994). Cyanobacteria are capable of carrying out both aerobic photosynthesis and anaerobic photosynthesis (Carr and Whitton 1982). The photosynthetic organ of cyanobacteria is similar to that of other algae in terms of structures and mechanisms. Nevertheless, this organ in cyanobacteria is characterized by a variety of forms and metabolisms, and determines the ecological roles of the organisms (Carr and Whitton 1982). Many cyanobacterial species are diazotrophic. They can assimilate nitrogen from the atmosphere (Ressom et al. 1994). In this case, either nitrogen fixation is carried out in heterocysts and photosynthesis occurs in vegetative cells, or alternate cycles of photosynthesis and nitrogen fixation are carried out (Rai 1990). Most cyanobacterial species have some characteristics that enable them to grow in widely fluctuating environments, including a capability to photosynthesize under low light conditions, a low energy requirement, buoyancy regulation mechanisms, a capability of withstanding long periods of extreme conditions, an improved capability for phosphorus uptake, and a mechanism for the efficient uptake of iron (Rai 1990).

Cyanobacteria generally reproduce by fission, budding, trichome breakage, hormogonia formation, or akinetic germination. Cyanobacteria can also reproduce by conjugation, transformation, and transduction (Rai 1990). In the right conditions, cyanobacteria can reproduce quite rapidly and so lead to the formation of a bloom. Replication times usually vary within and between species and depend on environmental factors such as temperature, light, and nutrient availability. The mean doubling time for several species was estimated to range from 21 hr to 14.7 d. The average for bloom-forming species is about 2 d under optimal conditions, and a bloom only persists for approximately 5–7 days (Ressom et al. 1994).

C. Factors Influencing the Development and Growth of Cyanobacteria

Environmental factors such as nutrients, light, and temperature are believed to have an influence on the formation of cyanobacterial blooms. Cyanobacterial blooms are likely to occur in both deep and shallow water bodies (Johnstone 1994). There is, however, disagreement about the factors that stimulate cyanobacterial blooms in Australia. Most Australian water managers and scientists believe that cyanobacterial blooms occur because of anthropogenic input of phosphorus to surface waters (Jones 1994). However, Jones (1994) postulated that the main cause of excessive growth of cyanobacteria in Australian inland waters was not excessive nutrient availability in the water but prolonged physical stability of the water column. He proposes that the cyanobacterial blooms are a "natural consequence of impounding water in a hot, dry country" (Jones 1994).

Although cyanobacterial blooms may be a natural phenomena in Australia, it is generally agreed that the frequency and intensity of their occurrence can be influenced by human activities. Disposal of nutrients into watercourses from sewage treatment plants, nutrient-rich urban runoff, intensive agricultural industries such as cattle feedlots and piggeries, and extensive agricultural practice can result in excessive aqueous nutrient levels from fertilizers and stock wastes (QWQTF 1992). It has been suggested that blooms are caused by a complex interaction of high concentrations of nutrients, sunlight, warm temperature, turbidity, pH, conductivity, salinity, carbon availability, and slow-flowing or stagnant water (Jones 1994).

III. Toxin Production by Freshwater Cyanobacteria

Freshwater cyanobacteria are capable of producing a wide range of toxic compounds as secondary metabolites, including neurotoxins [anatoxin-a, anatoxin-a(s), saxitoxin, and neosaxitoxin], hepatotoxins (microcystins, cylindrospermopsin, and nodularin), and lipopolysaccharide endotoxins (Carmichael 1988). All the common bloom- and scum-forming cyanobacterial genera have the potential to produce toxins (Gorham and Carmichael 1988; Codd et al. 1989, 1994). In systematic surveys around the world, 25%–70% of cyanobacterial water blooms were found to be potentially toxic (Lahti 1997). Of some 50 genera of freshwater cyanobacteria, at least 7 (*Anabaena, Aphanizomenon, Coelosphaerium, Gloetri-*

chia, Microcystis, Nodularia, and Nostoc) contain some toxic species (Lukac and Aegerter 1993).

A. Toxin Production by Cyanobacteria and Related Factors

Toxin production of cyanobacteria appears highly variable, both within and between blooms (Codd and Bell 1985). Toxin production and potency can also vary with time for an individual bloom. The factors contributing to toxin production in cyanobacteria are poorly understood (Ressom et al. 1994). There is evidence that toxin production and accumulation are correlated with growth, and so toxin production decreases rapidly when growth ceases. Indeed, toxin production tends to increase during the exponential growth phase and decrease gradually during the stationary phase (Watanabe and Oishi 1985). In addition, the rapid decline in toxicity at the cessation of the logarithmic growth phase is believed to be caused by a high rate of decomposition (Ressom et al. 1994).

Ressom et al. (1994) reported on a study conducted in Hartbeesport Dam, South Africa, that concentration of toxins in *Microcystis aeruginosa* was positively correlated with primary production per unit of chlorophyll *a*, solar radiation, surface water temperature, pH, and percentage oxygen saturation, but negatively correlated with surface water organic and inorganic nutrient concentration.

One of the factors influencing toxin production that has attracted the attention of many scientists is light intensity, but there is disagreement regarding this factor. Watanabe and Oishi (1985) considered light intensity as the primary important factor for the toxin production of cyanobacteria. They believed that production of the toxins was suppressed by low light intensity. A fourfold increase in toxicity of *M. aeruginosa* in culture was observed when the light intensity was increased from 7.5 to 200 $\mu E\ m^{-2}\ S^{-1}$ (Watanabe and Oishi 1985). In contrast, Van der Westhuizen and Eloff (1985) found a decline in toxicity at this high light level. In another study, relatively little variation in toxicity of *M. aeruginosa* was observed when light intensity varied between 5 and 50 $\mu E\ m^{-2}\ S^{-1}$ (Codd and Poon 1988). Sivonen (1990) also inferred that light intensity markedly influenced toxin production by *Oscillatoria* strains, but found that low light intensity led to higher toxin production, compared to that produced at high light intensity. However, the leakage of toxins from the cells was higher at high light intensities. The above results have demonstrated that the effects of light on toxin production are not clear and seem to be variable between different strains and species (Ressom et al. 1994).

Temperature has also been evaluated and appears to have an influence on toxin production by cyanobacteria. Most strains of *M. aeruginosa* produced less amounts of toxin at the upper and lower limits of temperature optima for that species (Gorham 1964; Van der Westhuizen and Eloff 1983; Codd and Poon 1988; Ressom et al. 1994). Watanabe and Oishi (1985) found that toxin production of cyanobacteria was reduced at high temperature (30°C). The study of Sivonen (1990) also showed that toxin production of *Oscillatoria* strains was lowest at 30°C. As well, Runnegar et al. (1983) reported a fourfold increase in toxicity of *Microcystis* isolates for cultures grown at 18°C as compared to cultures at 29°C.

Apart from light intensity and temperature, the pH of the aquatic environment also seems to have effects on toxin production of cyanobacterial species. Van der Westhuizen and Eloff (1983) found that the toxicity of *M. aeruginosa* was greatest at the higher or lower pH at which the growth rate of this species was greatest.

Another environmental factor believed to have an effect on toxin production of cyanobacteria is nutrient concentration. Although Runnegar et al. (1983) found that nutrient levels did not influence toxin production of *M. aeruginosa*, other studies have reported different results. Codd and Poon (1988) observed an approximate 10-fold decrease in toxicity of *Microcystis* after removal of either nitrogen or inorganic carbon from the cultures, but no changes were caused by the removal of phosphorus from culture media. Sivonen (1990) found a correlation between high toxicity and high nitrogen concentration in the culture, and that toxin production was also related to the concentration of phosphorus between 0.4 and 5.5 mg/L phosphorus. However, no additional effect was observed at higher concentrations of phosphorus in the cultures (Sivonen 1990).

In the study of the influence of trace metals on toxin production of *M. aeruginosa*, Lukac and Aegerter (1993) found that only iron and zinc had effects on toxin production of *M. aeruginosa*. Zinc seemed to be required for optimal growth as well as toxin production. In contrast, absence or presence of a low level of iron (≤ 2.5 μM) resulted in an increase in toxin production, coupled with a decrease in growth rate, in *M. aeruginosa*.

Currently, there is little knowledge of the effects of the metabolic or genetic processes that regulate toxin production of cyanobacteria. Because toxin production varies among different strains of the same species (Carmichael and Gorham 1981; Eloff and Van der Westhuizen 1981), some scientists have suspected that toxin production may be under genetic control (Ressom et al. 1994). Vackeria et al. (1985) and Codd and Poon (1988) speculated that toxin genes may be contained in plasmids that are present in many cyanobacterial species. A few studies have been undertaken to clarify the function of plasmids on toxin production of cyanobacteria. Nevertheless, findings of these studies were unable to provide a definitive answer to the question of whether plasmids are involved (either directly or indirectly) in toxin production.

In conclusion, factors that may influence toxin production of cyanobacteria include light intensity, temperature, nutrient level, pH, and trace metals (e.g., iron, zinc). Even though there is disagreement in regard to the precise effects of these factors, it is clear that all of them play a role in the formation and development of cyanobacterial blooms, which constitute a primary source of toxins in water supplies.

B. Neurotoxin-Producing Species

Neurotoxins are produced by species and strains of the genera *Anabaena* (Carmichael et al. 1990), *Aphanizomenon* (Mahmood and Carmichael 1986), *Nostoc* (Davidson 1959), and *Oscillatoria* (Sivonen et al. 1989; Skulberg et al. 1992; Edwards et al. 1992). Various cyanobacterial strains producing different types of neurotoxins are illustrated in Table 1.

Table 1. Neurotoxin-producing cyanobacteria.

Neurotoxins	Strains or species	References
Anatoxin-a	*Anabaena flos-aquae*	Huber 1972; Devlin et al. 1977; Carmichael 1992b
	Anabaena spiroides	Carmichael 1992b
	Anabaena circinalis	Sivonen et al. 1989
	Oscillatoria	Edwards et al. 1992
	Aphanizomenon	Codd et al. 1997
	Microcystis	Codd et al. 1997
Anatoxin-a(s)	*Anabaena flos-aquae*	Mahmood and Carmichael 1986; Henriksen et al. 1997
	Anabaena lemmermannii	Onodera et al. 1997
Saxitoxin and neosaxitoxin	*Aphanizomenon flos-aquae*	Sawyer et al. 1968
	Anabaena circinalis	Humpage et al. 1994; Negri et al. 1994

C. Hepatotoxin-Producing Species

A range of freshwater cyanobacterial species can produce hepatotoxins (either microcystins, nodularin or cylindrospermopsin). Microcystins were first reported in strain NRC-1 (SS-17) of *M. aeruginosa* in 1956 (Bishop et al. 1959). Some *M. aeruginosa* strains such as NRC-1 (SS-17) produce a single toxin, while others are capable of producing several toxins. For example, strain WR 70 can produce four microcystins, and other strains can produce as many as six types of microcystins (Lembi and Waaland 1988). Luukkainen et al. (1993) isolated eight microcystins from 13 freshwater *O. agardhii* strains collected from different Finnish lakes. All strains produced from two to five microcystins (Luukkainen et al. 1993). Nowadays, microcystins are believed to be secondary metabolites of a range of cyanobacterial species, including *Microcystis, Anabaena, Nodularia, Nostoc,* and *Oscillatoria* (Ressom et al. 1994). According to Lahti (1997), *M. aeruginosa, M. viridis, Anabaena flosaquae, Anabaena* spp., *Oscillatoria agardhii, Nostoc* sp., *and Anabaenopsis milleri* all have strains capable of producing microcystins.

Another toxin, cylindrospermopsin, has been isolated and characterized from the cyanobacterium *Cylindrospermopsis raciborskii* in Australia (Hawkins et al. 1985; Chiswell et al. 1997) and *Umezakia natans* at Lake Mikata in Japan (Ohtani and Moore 1992; Harada et al. 1994a). Recently, cylindrospermopsin was also isolated from *Aphanizomenon ovalisporum* in Queensland, Australia (Shaw et al. 1999). Unlike microcystins or cylindrospermopsin, nodularin is produced only by the cyanobacterium *Nodularia spumigena* (Rinehart et al. 1988). Species capable of producing hepatotoxins are summarized in Table 2.

Table 2. Hepatotoxin-producing cyanobacteria.

Neurotoxins	Strains or species	References
Microcystins	*Microcystis aeruginosa*	Lahti 1997
	Microcystis viridi,	
	Anabaena flos-aquae	
	Anabaena spp.	
	Oscillatoria agardhii	
	Nostoc sp.	
	Anabaenopsis milleri	
Nodularin	*Nodualaria spumigena*	Rinehart et al. 1988
Cylindrospermopsin	*Cylindrospermopsis raciborskii*	Hawkins et al. 1985
	Umezakia natans	Harada et al. 1994a
	Aphanizomenon ovalisporum	Shaw et al. 1999

D. Species Producing Lipopolysaccharide Endotoxins

Lipopolysaccharide (LPS) endotoxins are a component of the outer cell walls of many cyanobacteria (Ressom et al. 1994). These toxins have been isolated and characterized from many freshwater cyanobacterial species, including *Schizothrix calcicola, Anabaena flos-aquae, Oscillatoria tenuis,* and *Oscillatoria brevis,* as well as *Microcystis aeruginosa, Anacystis nidulans,* and *Anabaena variabilis* (Kaya 1996).

IV. Chemical Nature of Freshwater Cyanobacterial Toxins

A. Chemical Nature of Neurotoxins

1. *Anatoxin-a.* Anatoxin-a is a tropane-related alkaloid toxin. It is a bicyclic secondary amine, 2-acetyl-9 azabicyclo [4.2.1] non-2-ene (Huber 1972; Devlin et al. 1977), with molecular weight of 165 daltons (Carmichael 1992b). Anatoxin-a is synthesized in the cell from ornithine via putrescine with the participation of the enzyme ornithine decarboxylase (Gallon et al. 1990). Because of its small size, anatoxin-a is rapidly absorbed when taken by the oral route. It mimics the effects of acetylcholine and acts at both the nicotinic and muscarinic receptors (Spivak et al. 1980; Aronstam and Witkop 1981). Anatoxin-a is not susceptible to enzymatic hydrolysis by cholinesterase because it does not possess an ester moiety (Kofuji et al. 1990). Anatoxin-a in the protonated form is more stable than the free base. It has a pK_a of 9.4 and exists in the protonated form at physiological pH (Koskinen and Rapoport 1985). The structure of anatoxin-a is presented in Fig. 1.

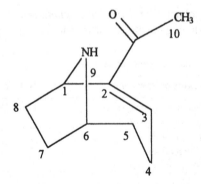

Fig. 1. Structure of anatoxin-a. (After Carmichael and Falconer 1993.)

2. *Anatoxin-a(s)*.   The chemical structure of anatoxin-a(s) is unrelated to anatoxin-a. It is a unique *N*-hydroxyguanidine methyl phosphate ester with molecular
weight of 252 daltons (Matsunaga et al. 1989). The structure is illustrated in Fig.
2.

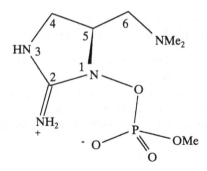

Fig. 2. Structure of anatoxin-a(s). (After Carmichael and Falconer 1993.)

3. *Saxitoxin and Neosaxitoxin*.   Saxitoxin and neosaxitoxin have a common
basic structure, but are unrelated to the other neurotoxins (Ressom et al. 1994).
They have been shown to be similar to known toxins produced by marine
dinoflagellate algae. Saxitoxin and neosaxitoxin are unique tricyclic molecules
with hydropurine rings. The structure of these toxins is shown in Fig. 3.

B. Chemical Nature of Hepatotoxins

1. *Microcystins*. Microcystins, which were first isolated from the Rideau River in
Ottawa, are cyclic heptapeptides with a molecular weight of about 1000 daltons
(Kuiper-Goodman et al. 1994). At least 50 variants of microcystins are known to
date (Bell and Codd 1996). The structures of microcystins were first presented by
Botes et al. (1982), and since then a number of structural variants have been

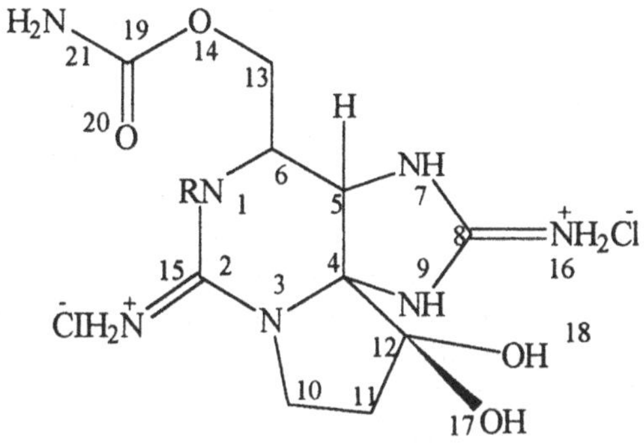

Fig. 3. Structure of saxitoxin and neosaxitoxin (R=H, saxitoxin; R=OH, neosaxitoxin). (After Carmichael and Falconer 1993.)

reported. According to Carmichael and Mahmood (1984), most microcystins contain β-methyl aspartic acid, glutamic acid, and alanine with methylamine attached to glutamic acid.

Microcystins contain seven amino acid peptides. These are homeometric peptides. Microcystins contain three invariant D-amino acids (Ala-erythro-β-methyl Asp, and Glu) that are highly 'conserved' (Botes et al. 1982; Ressom et al. 1994), two variant L-amino acids, N-methyl dehydroalanine, and a β-amino acid (Botes et al. 1982; Siegelman et al. 1984). The general structure of the microcystins is cyclo (-D-Ala-X-D-Me-Asp-Z-Adda-D-Glu-Mdha-) where X and Z are variable L-amino acids, D-Me-Asp is D-erythro-β-methylaspartic acid, Mdha is N-methyldehydroalanine, and Adda is (2S, 3S, 8S, 9S)-3-amino-9-methoxy-2,6,8,-trimethyl-10-phenyldeca-4,6,-dienoic acid (Botes et al. 1984; Luukkainen et al. 1993). The Adda chain determines the biological activity of the toxins (Carmichael 1992b; Stotts et al. 1993). In fact, changes in both Adda chain and D-glutamic acid reduced the toxicity of microcystins (Stotts et al. 1993). Therefore, not only Adda but also D-glutamic acid may have a significant role in determining the toxicity of microcystins.

All microcystins contain L amino acids at two 'nonconserved' positions in the molecule (Botes et al. 1982; Ressom et al. 1994). The most common L-amino acid X is leucine (L), arginine (R), or tyrosine (Y) (Carmichael 1988). Other amino acids can also be found in variants of microcystins, including homotyrosine (Hty) (Harada et al. 1991; Namikoshi et al. 1992), alanine (A), phenylalanin (F), homophenylalanin (Hph), methionine-S-oxide [M(O)], and trytophan (W) (Luukkainen et al. 1993). In the structure of microcystins, Z is arginine (R) or alanine (A) (Carmichael 1988). It might be also amino-isobutyric acid (Aba) (Gathercole and Thiel 1987), homoarginine (Har) (Namikoshi et al. 1990; Sivonen et al. 1991; Luukkainen et al. 1993), or methionine S-oxide [M(O)] (Botes et al. 1985).

Different types of microcystins are formed by substitutions in amino acid sites, by methylation or demethylation of the molecule, or by variations in the structure of the Adda chain (Bell and Codd 1996). Common variations also include demethylation of D-MeAsp (i.e., D-Asp) or Mdda (i.e., dehydroalanine [Dha]) (Carmichael 1981; Kristnamurthy et al. 1989; Meriluoto et al. 1989; Harada et al. 1991; Namikoshi et al. 1990; Luukkainen et al. 1993). In some microcystins, Mdha is replaced by L-serine (Namikoshi et al. 1990; Luukkainen et al. 1993) and D-Ala is replaced by D-serine (Luukkainen et al. 1993).

The molecular weight of microcystins is estimated to range from 500 to 4000 daltons. They are the largest of the cyanobacterial toxins (Codd and Bell 1985). However, most known microcystins have molecular weights between 909 and 1115 daltons (Steffensen and Nicholson 1994). The formula and molecular weight of common microcystins are presented in Table 3, and the structure of three common microcystins is illustrated in Fig. 4.

Table 3. Name, molecular weight, and formula of common microcystins.

Microcystins	Molecular weight	Formula
Microcystin-LA	909	$C_{46}H_{67}N_7O_{12}$
Microcystin-Laba	923	$C_{47}H_{69}N_7O_{12}$
Microcystin-AR	952	$C_{49}H_{68}N_{10}O_{12}$
Microcystin-YA	959	$C_{49}H_{65}N_7O_{13}$
[D-Asp[3], Dha[7]]microcystin-LR	966	$C_{47}H_{70}N_{10}O_{12}$
[D-Asp[3]]microcystin-LR	980	$C_{48}H_{72}N_{10}O_{12}$
[Dha[7]]microcystin-LR	980	$C_{48}H_{72}N_{10}O_{12}$
[DMAdda[5]]microcystin-LR	980	$C_{48}H_{72}N_{10}O_{12}$
Microcystin-LF	985	$C_{52}H_{71}N_7O_{12}$
Microcystin-LR	994	$C_{49}H_{74}N_{10}O_{12}$
[D-Asp[3], D-Glu(OCH$_3$)[6]]microcystin-LR	994	$C_{49}H_{74}N_{10}O_{12}$
[(6Z)-Adda[5]]microcystin-LR	994	$C_{49}H_{74}N_{10}O_{12}$
[L-ser[7]]microcystin-LR	998	$C_{48}H_{74}N_{10}O_{13}$
Microcystin-LY	1001	$C_{52}H_{71}N_7O_{13}$
Microcystin-HilR	1008	$C_{50}H_{76}N_{10}O_{12}$
[D-Asp[3],ADMAdda[5]],microcystin-LR	1008	$C_{49}H_{72}N_{10}O_{13}$
[D-Glu-OCH$_3$[6]]microcystin-LR	1008	$C_{50}H_{76}N_{10}O_{12}$
[D-Asp[3],Dha[7]]microcystin-PR	1009	$C_{47}H_{71}N_{13}O_{12}$
[L-MeSer[7]]microcystin-LR	1012	$C_{49}H_{76}N_{10}O_{13}$
[Dha[7]]microcystin-LR	1014	$C_{51}H_{70}N_{10}O_{12}$

Table 3. (Continued).

Microcystins	Molecular weight	Formula
[ADMAdda^5]microcystin-LR	1022	$C_{50}H_{74}N_{10}O_{13}$
[D-Asp^3,ADMAdda^5]microcystin-Lhar	1022	$C_{50}H_{74}N_{10}O_{13}$
[D-Asp^3]microcystin-RR	1023	$C_{48}H_{73}N_{13}O_{12}$
[Dha^7]microcystin-RR	1023	$C_{48}H_{73}N_{13}O_{12}$
Microcystin-FR	1028	$C_{52}H_{72}N_{10}O_{12}$
Microcystin-M(O)R	1028	$C_{48}H_{72}N_{10}O_{13}S$
[Dha^7]microcystin-HphR	1028	$C_{52}H_{72}N_{10}O_{12}$
[D-Asp^3, Dha^7]microcystin-HtyR	1030	$C_{51}H_{70}N_{10}O_{13}$
[Dha^7]microcystin-YR	1030	$C_{51}H_{70}N_{10}O_{13}$
[D-Asp^3]microcystin-YR	1030	$C_{51}H_{70}N_{10}O_{13}$
Microcystin-YM(O)	1035	$C_{51}H_{69}N_7O_{14}S$
[ADMAdda^5]microcystin-Lhar	1036	$C_{51}H_{76}N_{10}O_{13}$
Microcystin-RR	1037	$C_{49}H_{75}N_{13}O_{12}$
[(6Z)-Adda^5]microcystin-RR	1037	$C_{49}H_{75}N_{13}O_{12}$
[D-ser^1,ADMAdda^5]microcystin-LR	1038	$C_{50}H_{74}N_{10}O_{14}$
[ADMAdda^5 ,MeSer^7]microcystin-LR	1040	$C_{50}H_{76}N_{10}O_{14}$
[L-ser^7]microcystin-RR	1041	$C_{48}H_{75}N_{13}O_{13}$
[D-Asp^3,MeSer^7]microcystin-RR	1041	$C_{48}H_{75}N_{13}O_{13}$
Microcystin-YR	1044	$C_{52}H_{72}N_{10}O_{13}$
[D-Asp^3]microcystin-HtyR	1044	$C_{52}H_{72}N_{10}O_{13}$
[Dha^7]microcystin-HtyR	1044	$C_{51}H_{72}N_{10}O_{13}$
Microcystin-(H$_4$)YR	1048	$C_{52}H_{76}N_{10}O_{13}$
[D-Glu-OC$_2$H$_3$(CH$_3$)OH^6]microcystin-LR	1052	$C_{52}H_{80}N_{10}O_{13}$
Microcystin-HtyR	1058	$C_{53}H_{74}N_{10}O_{13}$
[L-Ser^7]microcystin-HtyR	1062	$C_{52}H_{74}N_{10}O_{14}$
Microcystin-WR	1067	$C_{54}H_{73}N_{11}O_{12}$
[L-MeLan^7]microcystin-LR	1115	$C_{52}H_{81}N_{11}O_{14}S$

Aba, aminoisobutyric acid; ADMAdda, *O*-acetyl-*O*-demethylAda; Dha, dehydroalanine; DMAda, *O*-demethylAdda; Har, hormoarginine; Hil, homoisoleucine; Hph, homophenylalanine; Hty, homo-tyrosine; MeLan, N-methyl-lanthionine; M(O), methionine-S-oxide; MeSer, N-methylserine; (6Z)-Adda, steroisomer of Adda at the Δ^6 double bond.
Adapted from Carmichael and Falconer (1993); Steffensen and Nicholson (1994).

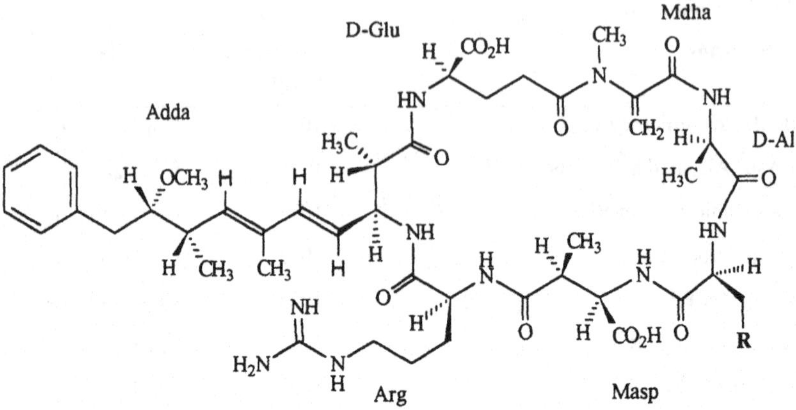

$$\text{Microcystin LR} \quad R = CH(CH_3)_2$$
$$\text{Microcystin RR} \quad R = CH_2CH_2NHC(NH_2) = NH$$
$$\text{Microcystin YR} \quad R = C_6H_4\text{-}p\text{-}OH$$

Fig. 4. Structures of some common microcystins. (Adapted from Sadler 1994.)

2. *Cylindrospermopsin*. Cylindrospermopsin is a tricyclic alkaloid, possessing 'a tricyclic guanidine moiety combined with hydroxymethylurocil' (Ohtani and Moore 1992). It has a molecular weight of 415 daltons and is believed to be derived from a polyketide that uses an amino acid-derived starter unit such as gly-cocyamine or 4-guanidino-3-oxybutyric acid (Moore et al. 1993). The structure of cylindrospermopsin is given in Fig. 5.

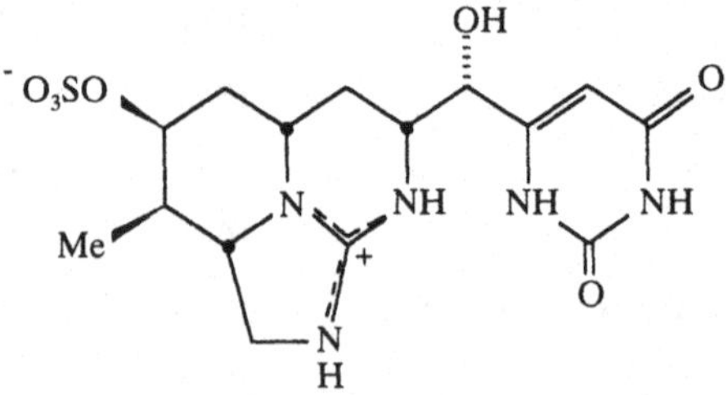

Fig. 5. Structure of cylindrospermopsin. (Adapted from Carmichael and Falconer 1993.)

3. *Nodularin*. The structure of nodularin is similar to that of microcystin. It is a monocyclic pentapeptide with a molecular weight of 824 daltons. Nodularin contains the amino acids D-glutamic acid, D-*B*-methylaspartic acid, L-arginine, D-*N*-methyl-dehydrobutyrine, and Adda (Rinehart et al. 1988). Seven different nodu-larins have been isolated from *Nodularia spumigena* blooms (Lahti 1997). The structure of nodularin is illustrated in Fig. 6.

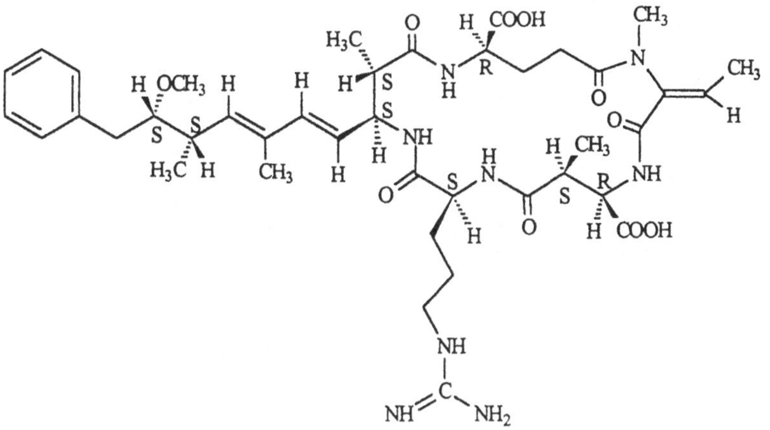

Fig. 6. Structure of nodularin. (After Beasley et al. 1989.)

C. Chemical Nature of Lipopolysaccharide Endotoxins

Lipopolysaccharide (LPS) endotoxins consist of three components, namely, lipid A in the innermost moiety, a core oligosaccharide in the middle moiety, and the O-specific polysaccharide in the outermost moiety (Jann and Jann 1984). The lipid A component of cyanobacterial LPS endotoxins is different and more variable than bacterial LPS (Mayer and Weckesser 1984). The toxins contain glucose, xylose, mannose, and rhamnose. The sugar and fatty acid composition may vary among the cyanobacterial species and are different from those of Enterobacteriaceae (Kaya 1996).

LPS endotoxins isolated from *M. aeruginosa* show the general properties of endotoxins. LPS endotoxins from *Anacystis nidulans* contain the LPS-specific sugar, 3-deoxy-D-mannooctulosonic acid. However, these components are absent in LPS endotoxins produced by *Anabaena variabilis*. These LPS endotoxins also contain no heptoses. Unlike LPS of gram-negative bacteria, cyanobacterial LPS endotoxins generally contain small amounts of phosphate. No phosphate is contained in the toxins of *A. variabilis*. In contrast, a significant amount of phosphate can be found in LPS endotoxins isolated from *M. aeruginosa* strain 006 and NRC-1 (Kaya 1996). The structure of LPS endotoxins is presented in Fig. 7.

V. Distribution, Absorption, Biotransformation, and Bioaccumulation of Freshwater Cyanobacterial Toxins

A. Distribution of Toxins in the Environment

1. *Distribution of Neurotoxins in the Environment.* There are few reports on the fate of neurotoxins in the environment. Some studies (e.g., Smith and Lewis 1987; Kiviranta et al. 1991; Stevens and Krieger 1991; Ressom et al. 1994) reported that certain neurotoxins are relatively labile molecules which tend to

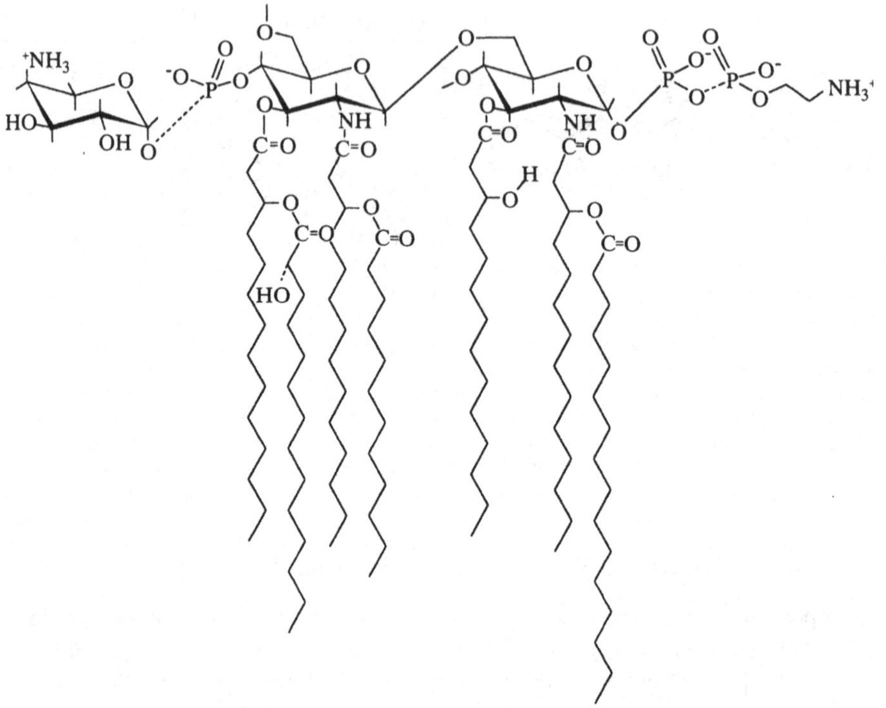

Fig. 7. Structure of LPS endotoxins. (From Kaya 1996.)

decompose rapidly under natural conditions to nontoxic breakdown products. Light and alkaline pH may cause a rapid degradation of anatoxin-a (Stevens and Krieger 1991), which is labile in sunlight. Under high light intensity, the degradation of anatoxin-a followed the first-order kinetics, with a half-life of 1–2 hr (Stevens and Krieger 1991). Degradation of anatoxin-a also appeared to depend on pH, and pH favorable for blue-green algal blooms may also be optimal for toxin degradation (Stevens and Krieger 1991).

Saxitoxin and neosaxitoxin are soluble in water and methanol. They are less soluble in ethanol and not soluble in acetone, ether, or chloroform. Saxitoxin and neosaxitoxin are stable in hot acid solution at pH 2–4, but become increasingly labile with increasing pH. The toxins are dialyzable, resistant to crystallization, hygroscopic, and decompose without melting (Jackim and Gentile 1968).

2. *Distribution of Hepatotoxins in the Environment.* Microcystins are the only hepatotoxins produced by freshwater cyanobacteria of which the distribution has been widely studied and reported in the scientific literature. Therefore, this part mainly focuses on microcystins. Microcystins are very stable and persistent (Ressom et al. 1994); they are stable at elevated temperatures, up to 300°C for extended periods (Lambert 1993), and are not susceptible to common proteolytic degradation (Falconer et al. 1986). Microcystins are also quite stable in sunlight, but can be decomposed or isomerized by sunlight in the presence of pigment.

Tsuji et al. (1995) observed that microcystin-LR was photodegraded with a half-life of about 5 d in the presence of 5 mg/mL of extractable cyanobacterial pigment. Microcystins are nonvolatile, dialyzable, and resistant to changes in pH (Ressom et al. 1994). The toxins are soluble in water, methanol, and ethanol, and insoluble in acetone, ether, chloroform, and benzene (Ressom et al. 1994). Microcystins produced by *Microcystis* and *Oscillatoria* exhibit lipophilic properties that are partly determined by the unsaturated side chain in Adda (Botes et al. 1985; Meriluoto and Ericksson 1988). Microcystins can diffuse through collodion, cellophane, and animal membranes (Ressom et al. 1994).

Microcystins are intracellular toxins (Hrudney et al. 1994b) and are only released into water by senescence or cell death or through water treatment processes such as prechlorination and algicide application (e.g., copper dosing) (Falconer 1993a; Jones and Orr 1994). Once released into water, microcystins can persist for a relatively long period (Watanabe et al. 1992) before being removed by biodegradation or photolysis (Hrudney et al. 1994b). Dissolved microcystin-LR has been found to persist longer than those in particulate materials (Lahti et al. 1997). Jones et al. (1995) reported that microcystins persisted in dry scum for up to 6 mon. After rewetting, microcystins were released from the crust material to the surrounding water (Jones et al. 1995). It was also observed that microcystins persisted for 2–3 mon in *Microcystis* cultures grown in water from the Vantaanjoki River in Finland (Ressom et al. 1994). Jones and Orr (1994) found microcystins up to 21 d following algicide treatment of a *Microcystis* bloom.

By adding microcystin-LR to a variety of natural surface waters, Jones (1990) observed that microcystin-LR persisted for 3 d to 3 wk before being degraded. Once begun, degradation was relatively rapid with more than 95% loss occurring within 3–4 d. The rate of degradation of microcystins may be increased if the water body has been previously exposed to microcystins (Jones 1990; Jones and Orr 1994; Jones et al. 1995). This increase in rate of degradation may be attributed to the availability of degradation bacteria. In natural waters, microcystins may be degraded by endemic bacterial populations (Jones and Orr 1994). The bacterium capable of degrading microcystins is a new *Sphingomonas* species, previously known as *Pseudomonas* species (Bourne et al. 1996). These microorganisms have been found in sewage effluent (Lam et al. 1995), lake sediments, and natural water bodies (Jones and Orr 1994; Rapala et al. 1994).

The breakdown of microcystin-LR may involve at least three enzymes and the formation of degradative intermediates (product A and B) (Bourne et al. 1996). These enzymes are assumed to be enzyme 1 (postulated as microcystinase, a novel metalloprotease), which catalyzes the conversion of microcystins to product A; enzyme 2, that is responsible for catalyzing the conversion of product A to product B; and enzyme 3, which catalyzes the breakdown of product B. Product A was assumed to be linearized (acryclo-) microcystin-LR as a result of the hydrolysis of microcystin-LR at either the Adda-Arg or Ala-Leu peptide bond. Product B was supposed to be a tetrapeptide (NH_2-Adda-iso-Glu-Mdha-Ala-OH) derived by further hydrolysis of product A. The process is illustrated in Fig. 8. Lam et al. (1995) reported that microcystin-LR biotransformation followed a

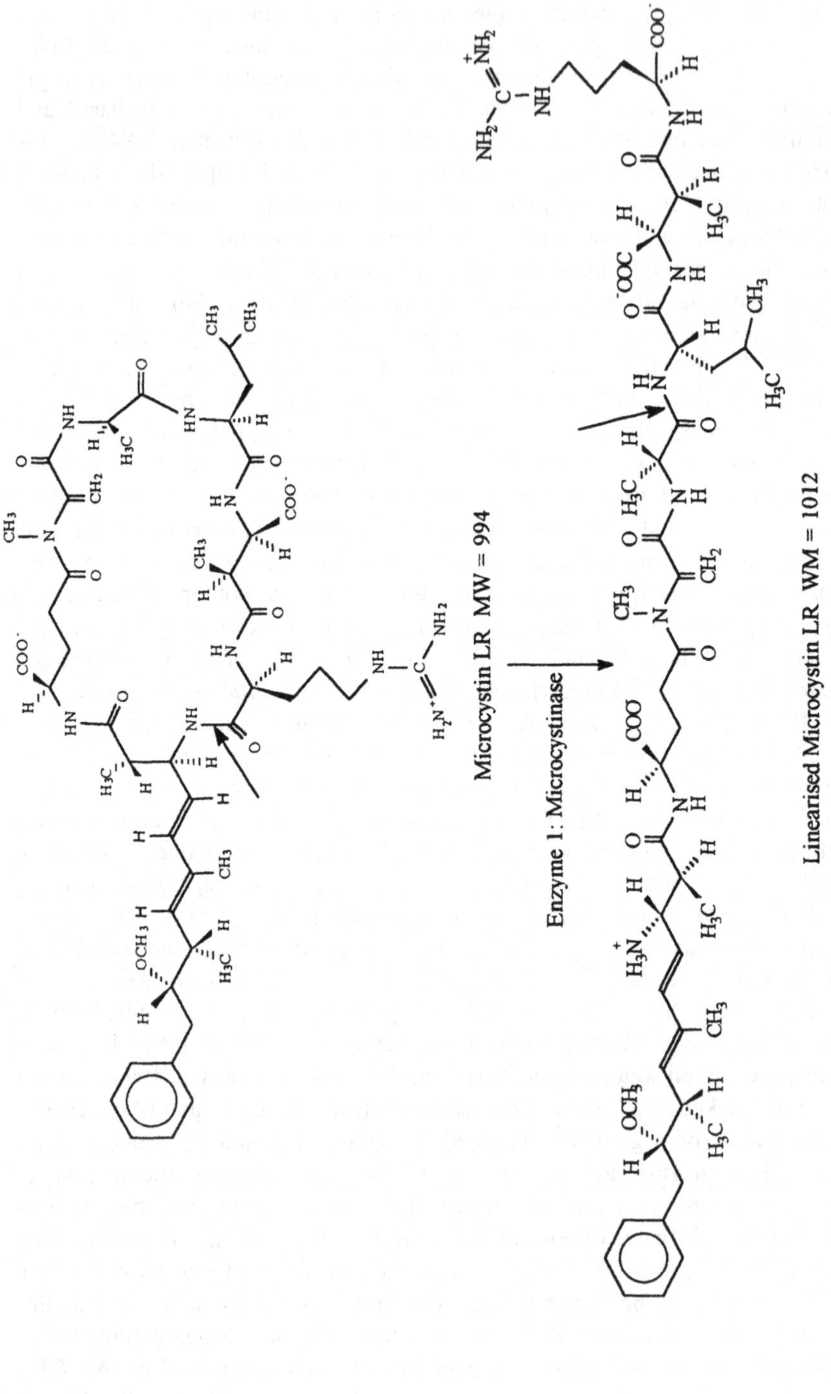

T. N. Duy et al.

Microcystin LR MW = 994

Enzyme 1: Microcystinase

Linearised Microcystin LR WM = 1012

Fig. 8. Proposed degradation pathway of microcystin-LR. *Small arrows*, sites of peptide hydrolysis; MW, molecular weight. (From Bourne et al. 1996.)

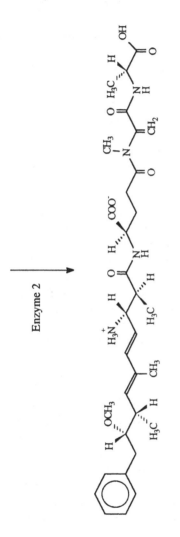

Enzyme 2

Tetra-Peptide (NH$_2$-Adda-isoGlu-Mdha-Ala-OH) MW = 614

Enzyme 3

Undetected Smaller Peptides and Amino Acids

Fig. 8. (Continued).

first-order decay; the first-order rate constant was 0.19 d^{-1}. The half-life of microcystin biotransformation in this study was estimated as 0.2–3.6 d. This half-life accords, to a certain extent, with a value of 3 d reported by Kenefick et al. (1993) and 0.5–2.8 d by Lam et al. (1995). The biotransformation rate of microcystin-LR by aerobic microorganisms in natural aquatic environments would be expected to be slower than the rates obtained under laboratory conditions (Lam et al. 1995).

In the study of the release and degradation of microcystin after algicide application, Jones and Orr (1994) found that the kinetics of microcystin degradation was biphasic. The first phase was a rapid process lasting for 3 d with 90%–95% loss of microcystins. The second phase was slower and continued until day 21, when a flood precluded further observation by the authors.

B. Uptake and Transformation of Microcystins in Mammals

Microcystins are unable to cross cell membranes and enter most tissues directly (Eriksson et al. 1990b). However, they can cross the ileum via the multispecific organic ion transport system and mainly enter hepatocytes (Kuiper-Goodman et al. 1994).

In mammals, ingested microcystins are transported through cell linings of the small intestine by the bile acid transport system and then transferred to the liver (Falconer et al. 1992). Microcystins can also enter hepatocytes by the special uptake mechanism of bile acid (Eriksson et al. 1990a; Lahti 1997). In the case of uptake of cyanobacterial cells, the toxins are released in the stomach and taken up preferentially by the ileum, which has higher levels of bile acid carriers (Dahlem et al. 1988). Then, the toxins are concentrated again by the hepatocytes via the bile acid carrier (Runnegar et al. 1981). The bile acid transporter can be found in the intestinal cells (Lambert 1993), as well as liver and kidney cells (Runnegar et al. 1981; Ericksson et al. 1990a; Hooser et al. 1991; Lambert 1993).

The liver appears to be the main target organ for both accumulation and excretion of microcystins (Falconer et al. 1986). Most animal experiments show the liver to be the target organ for microcystin-LR, accumulating approximately 50%–70% of the total dose. The other organs that may accumulate significant amounts of microcystin-LR are the intestine and kidney (Robinson et al. 1989; Meriluoto et al. 1990; Lambert 1993). Other organs appear to absorb only minor quantities of this toxin. In tissue distribution studies carried out with microcystin-LR, 50%–70% of the toxin was found in the liver, with another 7%–10% in the intestine, 1%–5% in the kidney, and the remainder distributed throughout the body (Hrudney et al. 1994a). Even though no study on the transport mechanism for microcystin-LR in the kidney has been reported, it is likely that transport may occur in the kidney as well because the kidney also has a bile acid transport system similar to the transporter of the intestinal cell in rats (Lambert 1993). The proportions of microcystin-LR compared with injected dose in organs are given in Table 4.

Microcystins are resistant to enzymatic hydrolysis (Runnegar and Falconer 1981) and have no "cleavage point for tryptic hydrolysis" (Ressom et al. 1994). Therefore, they are more resistant to degradation in tissues and may require

Table 4. Tissue distribution of microcystin-LR in mice and rats.

1: Tissue distribution of microcystin-LR in mice (%)

Liver	Intensive	Kidney	Carcass Organ	References
50–60	7	1	10	Robinson et al. 1989
70	10	3–5	10	Meriluoto et al. 1990

2: Tissue distribution of iodinated microcystin-LR in rats (%)

Liver	Kidney	Gut content	Urine	References
20	5	9	2	Falconer et al. 1986

excretion in bile in either its original form or after conjugation (Falconer et al. 1986). There is evidence for biliary excretion of microcystins from the liver, indicating the potential for enterohepatic recirculation of toxins (Falconer et al. 1986). The authors suggested that microcystins may leave hepatocytes via the bile because they rapidly appear in the duodenum. It is conceivable that microcystin-LR may be excreted through both liver and kidney routes (Lambert 1993). Microcystins and their breakdown products from the plasma are excreted through urine and feces. Plasma half-lives of microcystin-LR after iv administration were 0.8 and 6.9 min for the α- and β-phases of elimination, respectively (Falconer 1993b).

Falconer et al. (1986) found that the disappearance of microcystins in blood followed two phases. The first phase had a half-life of 2.1 min and the second phase 42 min (Falconer et al. 1986). Rates of microcystin clearance in the body decrease as the extent of liver damage increases (Steffensen and Nicholson 1994). It appears that different animal species vary in their ability to clear microcystins. For instance, Atlantic salmon seems to clear microcystin-LR at a higher rate than mice (Robinson et al. 1989).

C. Bioaccumulation of Microcystins

Bioaccumulation of cyanobacterial toxins in food chains may be of concern because it could have undesirable effects on public health and natural ecosystems. Even though there are only a few studies in this area, findings from these studies have shown that microcystins may be accumulated in freshwater mussels (Eriksson et al. 1989), freshwater clams (Prepas et al. 1997), and fish (Rabergh et al. 1991). Therefore, transfer of the toxins through the food web may be possible.

In an experiment on the freshwater mussel *Anadonta cygnea*, Eriksson et al. (1989) found that the mussels readily accumulated *Oscillatoria* toxin (microcystins). The mean toxin concentration was 70 µg/g dry weight after 15 d of exposure to 40–60 µg toxin/L. Total amount of toxin in individual mussels was 70–280 µg/mussel, and the highest toxin concentration was detected in the hepatopancreatic tissue, accounting for 40% of the total toxin content. The toxins were also detected in intestine, gonad, kidney, and connective tissue in minor quantities (Eriksson et al. 1989). Because *A. cygnea* can accumulate *Oscillatoria* toxins, Eriksson et al. (1989) speculated that bioaccumulation of the toxins or similar peptide toxins may occur in other filter feeders and, in turn, these contaminated organisms may contribute in transferring the toxins to higher trophic levels.

Likewise, bioaccumulation of microcystins was observed by Prepas et al. (1997). They found that clams bioconcentrated microcystin-LR to measurable concentrations in lakes where toxin concentration in the phytoplankton was undetectable. They also observed that higher microcystin concentration in phytoplankton resulted in a higher microcystin concentration in clams. The toxin was found in three body parts (visceral mass, gill, and muscle tissue) in relatively equal amounts. When clams were removed from the toxin source and fed nontoxic green algae, the biphasic elimination of microcystins occurred. Of the total accumulated microcystins, 71% was eliminated in the first 6 d, with 70%, 88%, and 69% loss in visceral mass, gill, and muscle tissue, respectively. No further toxin elimination was found in the visceral mass and gill tissue. Between day 6 and 21, however, losses from muscle tissue were recorded, increasing from 69% at day 6 to 81% at day 21. The 29% of the toxins that remained in the clams after 6 d did not change after 21 d (Prepas et al. 1997). Through clams, the toxins may be transferred through both aquatic and terrestrial food webs by the consumption of these contaminated animals.

In addition, Watanabe et al. (1992) observed that microcystin-LR, -RR, and -YR were accumulated in the zooplankton community of *Bosimina fatalis*, *Diaphanosoma brachyuru*, and *Cyclopsvicinus*. The amount of microcystins found in these zooplankton was relatively high (75-1387 µg/g dw). *B. fatalis* appeared to accumulate the largest amount of microcystins. Based on these findings, the authors suggested that microcystins may be transferred to higher trophic levels via *B. fatalis* (Watanabe et al. 1992). Microcystins have also been detected in mussels collected from the coastal waters of British Columbia (Chen et al. 1993) and in aquatic invertebrates fouling Atlantic salmon (*Salmo salar*) net pens (Kotak et al. 1996b), as well as in the livers of salmon with severe liver disease (Kent et al. 1988; Kent 1990; Kotak et al. 1996b). Even though the sources of microcystins accumulated in these marine organisms are unknown, these findings have demonstrated that microcystins can be accumulated by aquatic organisms.

Furthermore, microcystins have been found in phytoplankton, zooplankton, and gastropods (*Lymnaea stagnalis*, *Helisoma trivolvis*, and *Physa gynna*) (Kotak et al. 1996b), as well as in copepods and crab larvae (William et al. 1995). The presence and biomass of the phytoplankton *M. aeruginosa* may determine the occurrence of microcystin-LR in food web compartments (Kotak et al. 1996b). Watanabe et al. (1992) also detected three microcystin analogues (microcystin-LR, -YR, and -RR), with concentrations ranging from 75 to 1387 µg/g, in zooplankton samples collected from Lake Kasumiguara, Japan. These findings again suggest that microcystins can be accumulated and transferred though the food webs by animals feeding on contaminated organisms.

VI. Toxicity and Health Effects of Freshwater Cyanobacterial Toxins

A. Mechanisms of Toxicity

1. *Neurotoxins.* Currently, the known neurotoxins produced by freshwater cyano-bacteria include anatoxin-a, anatoxin-a(s), saxitoxin, and neosaxitoxin (Ressom et al. 1994). Their target is the neuromuscular system, and they can paralyze peripheral skeletal and respiratory muscles. Death can result from respiratory arrest, occurring within a few minutes to a few hours (Carmichael et al. 1985).

Anatoxin-a. Anatoxin-a is a potent nicotinic agonist that acts as a postsynaptic, depolarizing, neuromuscular blocking agent (Carmichael et al. 1990; Harada et al. 1994b). The cholinergic system is the primary target of this toxin (Mahmood and Carmichael 1986). The major effects of anatoxin-a are on the postsynaptic part of the neuromuscular junction. Anatoxin-a can also exhibit a presynaptic action that leads to a reduction in the frequency of miniature end-plate potential and in the quantal content of the end-plate potential (Carmichael et al. 1979).

Anatoxin-a(s). Anatoxin-a(s) is a cholinesterase (ChE) inhibitor, and this inhibi-tion is irreversible (Mahmood and Carmichael 1987). It was shown to inhibit *in vitro* electric eel acetylcholinesterase (AChE) and horse serum butyrylcholinest-erase (BUChE) by a mechanism similar to the organophosphate anticholinest-erases (Carmichael et al. 1990). Anatoxin-a(s) also sensitized the frog rectus abdominus muscle and chick biventer cerviers muscle to exogenous acetylcho-line. Anatoxin-a(s) has a higher affinity for human erythrocyte AChE than electric eel AChE (Carmichael et al. 1990). Thus, this toxin has a significant potential to poison humans.

Other Neurotoxins. Saxitoxin and neosaxitoxin are fast-acting neurotoxins that inhibit nerve connections by blocking sodium channels without affecting perme-ability to potassium transmembrane resting potential or membrane resistance (Adelman et al. 1982). These sodium channel-blocking agents inhibit transmis-sion of nervous impulses and can lead to death by respiratory arrest (Carmichael and Falconer 1993).

2. *Hepatotoxins.* Hepatotoxic freshwater cyanobacterial toxins involve the cyclic peptides (microcystins, nodularin) and the alkaloid toxin (cylindrospermopsin). Mechanisms of action for each toxin are described as follows.

Microcystins. Microcystins have effects on two kinds of cells, namely hepato-cytes and macrophages (Kaya 1996). In mammals, after consumption through the oral route, microcystins are transported through cell linings of the small intestine by the bile acid transport system to the liver (Falconer et al. 1992). Microcystins can also enter hepatocytes via the special uptake mechanism of bile acids (Eriks-son et al. 1990a). Once in hepatocytes, microcystins act as an inhibitor of protein phosphatase and activator of phospholipase A_2 and cyclooxygenase (Kaya 1996). Microcystins can cause cell deformation and the development of blebs (Lahti 1997). According to Eriksson et al. (1989) and Lahti (1997), this deformation is the consequence of breakdown of intermediate filaments of the cell cytoskeleton and changes in actin microfilaments. It has been hypothesized that microcystins interact with the cytoskeletal elements of the cell to cause the deformities observed both *in vitro* and *in vivo* and the disruption of the cellular architecture (Carmichael 1994).

Microcystins have been shown to be an inhibitor of serine/threonine protein phosphatase 1 and 2A (Honkanen et al. 1990; Lahti 1997). This inhibition leads to hyperphosphorylation of proteins associated with the cytoskeleton. Through inhibition of phosphatases, microcystins disturb the regulation of phosphorylation of subunit proteins on the disassembly of intermediate filaments in normal cell mitosis. Consequently, they cause cytoskeletal disintegration (Falconer 1993b). The rapid loss of sinusoidal architecture leads to a concentration of blood in the liver and hypovolemic shock (Runnegar and Falconer 1982; Lahti 1997). The inhibition of protein phosphatases 1 and 2A by microcystins occurs in the nanomolar concentration range (Nishiwaki-Matsushima et al. 1991).

In macrophages, microcystins cause production of tumor necrosis factor-α (TNF-α) and interleukin-1 (IL-1). These cytokineses will produce platelet-activating factor (PAF) and activate cyclooxygenase.

Nodularin. Nodularin is a peptide toxin, and the mechanism of toxicity of nodularin is quite similar to that of microcystins. First, the toxins enter the blood from the ileum via the bile acid carriers that convey the peptide toxins across the mucosa (Carmichael 1992b). Second, the toxins are transported preferentially into the hepatocytes via bile acid carriers (Runnegar et al. 1981; Dabholkar and Carmichael 1987; Meriluoto et al. 1990; Eriksson et al. 1990a). Third, the peptide toxins induce changes in the actinmicrofilaments and elements of the cell cytoskeleton, and so result in a dense aggregation of the microfilaments near the center of the cell (Runnegar and Falconer 1986; Eriksson et al. 1989; Hooser et al. 1991). The loss of cellular support may cause cells to roundup and in turn result in the destruction of the sinusoid endothelial cells. Eventually, destruction of the parenchymal cells and sinusoids of the liver cause lethal intrahepatic haemorrhage (within hours) or hepatic insufficiency (within days) (Carmichael 1994).

Cylindrospermopsin. Once consumed through the oral route, cylindrospermopsin can cause gastroenteritis through injury to the gut lining, hepatitis from injury to liver cells, renal malfunction from cell injury to the kidneys, and hemorrhage from blood vessel injury (Falconer 1994a). Terao et al. (1994) described four consecutive phases of pathomorphological changes induced by cylindrospermopsin in the liver. In the initial phase, ribosomes detach from the membranes of the rough-surfaced endoplasmic reticulum and accumulate into the cytoplasm of hepatocytes. Usually, this process is accompanied by the condensation and reduction in the size of nucleoli. The second phase often begins 24 hr after administration and is correlated with membrane proliferation. In this phase the amount of total P-450 is considerably decreased in the toxin-treated hepatic microsomes. The authors believed that marked proliferation of agranular membranes is due to lipid peroxidation caused by P-450. The third phase is represented by an accumulation of fat droplets in the central portion of hepatic lobules, probably induced by free radicals generated in the injury. Also, the last phase is characterized by severe liver necrosis (Terao et al. 1994).

3. *Lipopolysaccharide Endotoxins.* Freshwater cyanobacterial LPS endotoxins are suggested (but unproven) to cause gastrointestinal disturbances (Sykora and Keleti 1981; Martin et al. 1989). These toxins are less toxic to rodents than the endotoxins of enteric pathogens such as *Salmonella* (Bell and Codd 1996). Nevertheless, the toxicity mechanism of LPS endotoxins produced by cyanobacteria is still largely unknown.

B. Effects of Toxins Observed from Animal Tests

1. *Physiological Effects.*

Anatoxin-a. Animals administered with acute lethal doses of anatoxin-a experience staggering, muscle fasciculations, gasping, convulsions, and opistothonos (head bent over back, in birds), viscous salivation and lachrymation (in mice), chromodacryorrhea (bloody tears, in rat), urinary incontinence, muscular weakness, and defecation. Death from respiratory arrest usually occurs within a few minutes (Carmichael 1992a). However, anatoxin-a in amounts less than those causing acute signs does not have significant effects on the nervous system or any other toxic effects besides its specific neuromuscular depolarizing activity. Anatoxin-a also did not appear to be a potent teratogen in hamsters (Astrachan et al. 1980).

Anatoxin-a(s). Anatoxin-a(s) is considered to be an irreversible cholinesterase inhibitor (Carmichael and Falconer 1993). In pigs, anatoxin-a(s) can cause hypersalivation, mucoid nasal discharge, tremors and fasciculations, ataxia, diarrhea, and recumbency. Observed additional symptoms such as regurgitation of algae, dilatation of cutaneous vessels in the webbed feet, wing and leg paresis, opisthotonos, and clonic seizures preceding death were reported in ducks (Beasley et al. 1989).

In mice administered with anatoxin-a(s), Mahmood and Carmichael (1987) and Cook et al. (1988) found lachrymation, dyspnea, cyanotic oral mucous membrane, viscous mucoid hypersalivation, urination, defecation, and clonic seizures before death. Death often occurs from respiratory arrest. In rats, chromodacryorrhea (red-pigmented "bloody tear") was also observed (Mahmood and Carmichael 1987).

Microcystins. Microcystins can cause pallor, lowered blood pressure and body temperature, hyperglycemia and tachycardia, and death by respiratory failure in mammals. Hematocrit, lowering of hemoglobin concentration, and lower red blood cell (RBC) count and total serum protein were also observed (Carmichael 1981). In sheep challenged with microcystins, there was a marked elevation of serum concentrations of certain enzymes and bilirubin, and mild elevations of blood urea nitrogen and serum inorganic phosphorus together with a marked reduction in blood glucose, a mild neutrophilia with a marked left shift, and marked changes in coagulation parameters (Jackson et al. 1984). Some sheep showed mild depression, increase in heart rate, variation in respiration rate, cessation of ruminal sounds and cudding, muscle twitching of the ears and eyelids and face and limb muscles, and paddling accompanied by gasping respiration (Jackson et al. 1984). In mice, rats, and calves, microcystins cause a progression of muscle fasciculations, decreased movement, collapse, exaggerated abdominal breathing, cyanosis, convulsions, and death (Carmichael 1992a).

Most experimental or natural poisonings of animals have shown that the target organ of microcystins is the liver. Low concentration may cause deformation in isolated hepatocytes. The deformation often leads to a complete loss of microvilli and development of clusters of large blebs or projections from the cell membrane. Microcystins also cause marked deformation with a large bleb or bleb forming in isolated enterocytes (Falconer 1993b).

With acute lethal doses, microcystins may result in severe liver damage, characterized by massive hepatocyte necrosis (Jackson et al. 1984). Disintegration of the sinusoidal lining cell of the liver and disruption of the cell membrane of hepatocytes are observed (Falconer et al. 1981). Necrotic cells have eosinophilic, frequently disintegrating cytoplasm. Pyknosis and karyorrhexis are often observed in the nuclei of necrotic cells. Because microcystins damage the liver, they lead to changes in concentration of enzymes (Honkanen et al. 1990). For example, an increase in aspartate aminotransferase and lactate dehydrogenase is usually observed in animals challenged by microcystins (Jackson et al. 1984; Honkanen et al. 1990). Falconer et al. (1994) also found an increase in GGT (γ-glutamyl transpeptidase), ALP (alkaline phosphatase), and total bilirubin, but a decrease in plasma albumin (in the pig experiment). Death of animals intoxicated with microcystins is believed to result from intrahepatic hemorrhage and hypovolemic shock (Carmichael 1992a). Microcystins generally do not have effects on the intestine, heart, spleen, kidney, or stomach (Carmichael and Mahmood 1984). However, in some animal tests, the effects of toxins were occasionally observed in the lung and the kidney (Østensvik et al. 1981; Carmichael and Mahmood 1984; Jackson et al. 1984; Berg et al. 1988).

Nodularin. The physiological effects of nodularin on exposed organisms resemble those produced by microcystins. For example, nodularin can cause liver damage and tumor promotion (Runnegar et al. 1988). Unlike the microcystins, however, nodularin may act as a tumor initiator (Lahti 1997). Therefore, nodularin is suspected to be a new environmental carcinogen (Kaya 1996).

Cylindrospermopsin. Mice affected by cylindrospermopsin huddle and are anorexic and usually experience slight diarrhea. Deaths often occur approximately 6–9 hr after injection with an acute lethal dose. Death is preceded by a slow gasping respiration and occasional limb paddling (Hawkins et al. 1985). Cylindrospermopsin may cause histologically recognizable injury or death to hepatocytes near the central veins (Hawkins et al. 1985).

The main target organ of cylindrospermopsin is also the liver, but the toxin has effects on a range of organs such as the thymus, kidneys, and heart (Terao et al. 1994). It also appears to be a potent inhibitor of protein synthesis (Terao et al. 1994). Hawkins et al. (1985) also observed the effects of cylindrospermopsin on the liver, kidneys, lungs, and adrenal glands of test mice. It caused hepatocellular coagulative necrosis and formation of fibrin thrombi in the portal veins of the livers. A variable epithelial cell necrosis was observed in the kidney. Adrenal cortices showed variable scattered epithelial cell degeneration and necrosis. The small intestine also underwent congestion and edema (Hawkins et al. 1985). At low doses, cylindrospermopsin causes injury or death to hepatocytes near central veins. High doses of cylindrospermopsin lead to extensive hepatocyte necrosis (Hawkins et al. 1985).

LPS endotoxins. Exposure to LPS endotoxins of cyanobacteria may result in gastroenteritis as well as skin irritation, eye irritation, allergic reactions, hayfever-like symptoms, and asthma (Falconer 1994b). However, these symptoms have not been demonstrated in laboratory studies.

2. *Genotoxicity.* Runnegar and Falconer (1982) reported that purified *Microcystis* extract was nonmutagenic in the Ames bacterial mutation test. According to Repavich et al. (1990), toxic extracts from *Microcystis, Anabaena,* and *Gleotrichia* collected from Wisconsin lakes in 1986 were not mutagenic in *Salmonella typhimurium* strains TA98, TA100, and TA102.

Most studies of the genotoxicity of cyanobacterial toxins have shown that cyanobacterial toxins have no gene mutagenicity in prokaryotic cells, but that microcystins were clastogenic with erythropoietic cells *in vitro* (humans) and *in vivo* (rats) (Ressom et al. 1994). In an *in vitro* test using human lymphocytes, chromatid breaks caused by hepatotoxins, anatoxin-a(s), and neosaxitoxin were observed (Ressom et al. 1994). The potency of this clastogenic effect was reported to be greater than that observed for benzene and sodium arsenite (known human carcinogens) and similar to 3,4,3',4'-polychlorinated biphenyl (probable human carcinogen) (IARC 1987).

Kirpenko et al. (1981) believed that toxins extracted from *M. aeruginosa* could penetrate the placental barrier to have effects on embryo in pregnant rats. They found increased fetal death, internal hemorrhage in fetuses, and mild teratogenicity after injecting *Microcystis* extracts into pregnant rats for the first 19 d after conception. In addition, Falconer et al. (1988) observed a brain abnormality in offspring of mice receiving *Microcystis* extracts through pregnancy. The abnormality included a reduction in brain size with damage in the outer region of the hippocampus (Falconer et al. 1988). Collins et al. (1981) also believed cyanobacterial toxins were associated with a high human birth defect rate in a community consuming water in a reservoir contaminated by cyanobacterial blooms.

3. *Tumor Promotion.* Microcystins (microcystin-LR, -YR, -RR) and nodularin have been shown to be potent inhibitors of protein phosphatases type 1 (PP1) and type 2A (PP2A) (Adamson et al. 1989; Honkanen et al. 1990; MacKintosh et al. 1990; Matsushima et al. 1990; Yoshizawa et al. 1990).

Bagu et al. (1995) and Craig et al. (1996) proposed that the toxins have effects on PP1 and PP2A through two steps. The first step is characterized by a rapid and reversible binding of the microcystins, which leads to potent but reversible inhibition of catalytic activity. The second step is represented by the formation of an irreversible covalent linkage with the N-methyldehydroalanine (Mdha) residue in the microcystins and a nucleophilic site on the phosphatase. This step usually lasts for several hours (Bagu et al. 1995; Craig et al. 1996).

MacKintosh et al. (1990) and Yoshizawa et al. (1990) suggested that the binding of microcystins and nodularin to PP1 and 2A results in the hyperphosphorylation of cellular proteins and a breakdown in normal protein function. Hyperphos-phorylation, in turn, influences the cell cytoskeleton, which results in a transition to an apparently mitotic state (Falconer and Yeung 1992). Consequently, tumor development occurs because increased mitosis leads to accelerated tissue growth. Furthermore, microcystins cause loss of cell–cell contact and may reduce the nor-mal contact inhibition of cell replication in organs and lead to tumor growth (Fal-coner and Yeung 1992). Therefore, microcystins and nodularin are believed to be potent tumor promoters, and this hypothesis has been confirmed by *in vivo* inves-tigations (Falconer 1991; Nishiwaki-Matsushima et al. 1992).

Experiments in which mice were given microcystins in their drinking water showed increased weight of carcinogen-initiated skin tumors (Falconer and Buck-ley 1989; Falconer 1991). Direct intraperitoneal (ip) injection of microcystin-LR in rats caused a promotion of liver tumor cell growth after chemical initiation of the tumors (Nishiwaki-Matsushima et al. 1992). Even though microcystins are not currently believed to be carcinogens, liver tumors were observed occasionally in animals administered with microcystins. In the study of Falconer et al. (1988), for instance, two bronchogenic carcinomas, one abdominal carcinoma, and a tho-racic lymphosarcoma were observed among 71 female mice receiving a toxin concentration of 14.5 µg/L, compared to one uterine adenocarcinoma and one thoracic lymphosarcoma among 223 mice on lower or no toxin doses.

C. Effects of Toxins on Aquatic Organisms

Cyanobacterial toxins appear to be toxic to zooplankton species. Bioassays have shown that *Microcystis* toxins can affect *Daphnia pulicania* and reduce filtering rates and the survival of newborn daphnids (Bell and Codd 1996). By exposing *Daphnia magna* to saxitoxin and neosaxitoxin, Sasner et al. (1984) observed that the characteristic movements of the second antennae of these organisms were erratic or stopped, causing the animals to settle to the bottom. Saxitoxin and neosaxitoxin also reversibly blocked the conduction of action potential in the crayfish giant axons (Sasner et al. 1984). Both *Daphnia hyalina* and *Daphnia pulex* were affected by microcystins (Hrudney et al. 1994a). Microcystins and anatoxin-a have also been found to be toxic toward brine shrimp (Bell and Codd 1996; Lahti 1997). In addition, Penaloza et al. (1990) found that certain zoop-lankton such as rotifers, copepods, and cladocerans were killed by microcystins. Cyanobacterial toxins are also believed to have effects on mosquito larvae (Ily-aletdinova and Dubitskiy 1972). It has been demonstrated that cyanobacterial tox-ins could inhibit the development of house fly (*Musca domestica*) larvae and cause mortality of gypsy moth (*Ocneria dispar*) larvae (Ressom et al. 1994). They also reduced the size of fly larvae that survived and prevented hatching of their eggs (Ressom et al. 1994).

A kill of approximately 2000 brown trout in Loch Leven and the River Leven, Scotland, in 1992 was believed to have been caused by toxins of *Anabaena flos-aquae* (Codd 1994). In this incident, killed fish had liver damage character-ized by necrosis, cellular degeneration, condensation and disintegration of nuclei, edema, and focal areas of congestion. Gill damage was severe, involving widespread secondary lamellar opposition, numerous necrotic cells in the lamellar epithelium, and epithelial ballooning and edema (Codd 1994).

The death of 6 tonnes of fish in Kezar Lake, New Hampshire (U.S.) has been attributed to unknown toxins produced by *Aphanizomenon flos-aquae* (Humpage et al. 1994). According to Anderson et al. (1993), microcystin-LR or a closely related microcystin has a link to netpen liver disease (NLD) in Atlantic salmon. The authors conducted an experiment on healthy salmon and found that these salmon when exposed to microcystin-LR exhibited symptoms similar to those in fish suffering from NLD.

Even though Codd (1994) stated that the specific contribution of cyanobacterial toxins to fish mortalities in natural or control water bodies is unclear, the findings of Anderson et al. (1993) and other authors have indicated that cyanobacterial toxins may have effects on fish communities and other aquatic organisms. For instance, it was found that microcystins caused hepatocellular damage in carp *Cyprinus carpio* and rainbow trout *Oncorhynchus mykiss* under laboratory conditions (Anderson et al. 1993).

Likewise, Kotak et al. (1996a) observed that rainbow trout injected with microcystin-LR experienced a color change from light silvery green to a dark olive-green. The fish also lost swimming coordination and buoyancy control. Even though the fish appeared to be relatively tolerant to high microcystin-LR doses, deaths were recorded in the 1000 μg/kg bw dose group (Kotak et al. 1996a). The authors believed that death of treated rainbow trout was caused by a general hepatic failure resulting from massive hepatocyte necrosis. The pathological effects of microcystin-LR on rainbow trout involved widespread hepatocellular swelling and lysis of hepatocyte plasma membranes, causing liquefactive necrosis (organelles floating in a milieu of cellular debris). Lesions observed in the kidney involved coagulative tubular necrosis with a dilation of Bowman's spaces (Kotak et al. 1996a).

Gaete et al. (1994) also found that microcystins inhibited enzymes of the gill microsomal fraction of *Cyprinus carpio* involved in ion pumps and exchangers. Based on this finding, the authors suggested that the death of fish observed during algal blooms could be caused by a loss of the ability of the gills to maintain homeostasis of the internal medium. As well, Zambrano and Canelo (1996) observed that microcystin-LR and microcystin-LR-like toxin produced from *M. aeruginosa* inhibited the reaction of the Na^+–K^+ pump of the gill of carp (*Cyprinus carpio*). The authors believed that this inhibition might block the gills function and so disrupt the ion homeostasis of the internal medium. In turn, the impairment of gill activity may result in fish death.

D. Effects of Toxins on Wild and Domestic Animals

Microcystins are responsible for intermittent but repeated cases of poisoning in wild and domestic animals (Galey et al. 1986). Microcystin-LR has been shown to be toxic to various insect species such as the third-instar diamond-backed moth (*Plutella xylostella*), adult house fly (*Musca domestica*), and the third-instar cotton leafworm (*Spodoptera littoralis*) (Delaney and Wilkins 1995). The authors found that microcystin-LR has appreciable insect toxicity, comparable to that of malathion, carbofuran, and rotenone. The effects of cyanobacterial toxins on animals (both wild and domestic) involve neurotoxicosis and hepatotoxicosis. The symptoms of neurotoxicosis that are often observed in affected animals include a progression of muscle fasciculations, decreased movement, abdominal breathing, cyanosis, convulsions, and death. Opisthotonos (rigid "S"-shaped neck) is often observed in avian species (Carmichael and Falconer 1993).

The signs of hepatotoxicosis in animals include weakness, reluctance to move about, anorexia, pallor of the extremities and mucous membranes, and mental derangement (Carmichael and Falconer 1993). Death occurs within a few hours to a few days and is often preceded by coma, muscle tremors, and general distress (Galey et al. 1987). Death is believed to be the result of intrahepatic hemorrhage and hypovolemic shock (Falconer et al. 1981, 1988; Theiss et al. 1988).

Cyanobacterial toxins were attributed as the cause of death of cattle, sheep, pigs, horses, dogs, cats, monkeys, muskrats, squirrels, rhinoceros, fish, birds, and invertebrates (Carmichael and Falconer 1993; Bell and Codd 1996). The reported animal poisoning incidents associated with freshwater cyanobacterial toxins from 1878 to 1991 are given in Table 5; however, no doubt there are many more that remain unreported.

E. Effects of Toxins on Human Health

Outbreaks of human illnesses associated with freshwater cyanobacterial toxins have been reported in the scientific literature during the past 50 years. The first human illness was an outbreak of gastroenteritis associated with cyanobacterial toxins occurring in the Ohio River in 1931. Unfortunately, the species and toxin(s) responsible for the illnesses were not identified (Tisdale 1931; Veldee 1931). Since then, illnesses caused by the toxins have been documented around the world, including Australia, Europe, Africa, Asia, and North America. Recently, deaths of hemodialysis patients in Brazil from the presence of microcystins in dialysis water have been documented (Jochimsen et al. 1998). The possible health effects and incidents of human illnesses related to cyanobacterial toxins are summarized in Tables 6 and 7.

VII. Dose–Response Relationships

The relationship between dose of cyanobacterial toxins and response in intoxicated animals has been reported. Cyanobacterial toxins have been tested in a range of animals such as mice, rats, sheep, cats, fish, and guinea pigs. These animals were tested using different routes of administration: intraperitoneal (ip) route with single dose, ip route with multistage administration, oral route with single dose, oral route with multistage administration, dermal route, etc. The response in test animals varied with sex, age, test species, exposure route, exposure period, and types of toxins. In terms of exposure method, the ip route appears to be most toxic. Among cyanobacterial toxins, microcystins have been exten-

Table 5. Animal poisoning incidents associated with cyanobacterial toxins.

Year	Location	Animals affected	Species producing toxins
1878	Lake Alexandrina	Sheep, horses, dogs, pigs	*Nodularia spumigena*
1882–1884	Minnesota, USA	Cattle, horse, hogs	*Gleotrichia echinulata*
1990	Minnesota, USA	Several cattle	*Aphanizomenon flos-aquae*
1917–1918	Alberta, Canada	Hogs, horses, cattle, poultry, wild bird	"Blue-green algae"
1918	Minnesota, USA	1 sheep, 17 hogs , and about 50 chicken	*Coelospharium kuetzzingianum ,Anabaena flos-aquae*
1924	Ontario, Canada	About 20 cattle	*Anabaena*
1928	LakeVesijarvi, Finland	About 40 cattle	*Anabaena lemmemannii*
1930	Minnesota, USA	9 cattle	*Microcystis flos-aquae; Microcystis aeruginosa*
1933	Minnesota, USA	More than 21 sheep and chickens	*Microcystis flos-aquae*
1933	Minnesota, USA	45 turkeys, 4 ducks, 2 geese, cow, pigs, horses, poultry	*Microcystis flos-aquae; Anabaena flos-aqauae; Aphanizomemnon flos-aquae*
1933	Minnesota, USA	3 cattle	*Microcystis flos-aquae*
1934	Lake Juksa, USSR	Cats	Blue-green algal bloom
1939	Colorado, USA	4 ducks, wild birds, carps, snakes, salamanders, a calf	*Anabaena flos-aquae*
1913–1943	Free State and Transvaal, South Africa	Thousands of cattle, sheep, and other animals	*Microcystis toxica (= aeruginosa)*
1943	Montana, USA	Sheep	Algae, including blue-green algae

Table 5. (Continued).

Year	Location	Animals affected	Species producing toxins
1944–1945	Iowa, USA	37 hogs, 4 sheep, 2 cattle, 3 horses; several dogs, cats, squirrels, chickens, turkeys, and songbirds	*Anabaena flos-aquae*
1945	Manitoba, Canada	A horse, several calves, 2 pigs, and a cat	Water bloom
1945	Bermuda	Cattle deaths	*Aphanizomenon flos-aquae*
1946	North Dakota, USA	Cattle and deer	Hepatotoxins from water bloom
1948	Iowa, USA	Few dogs	*Anabaena flos-aquae*
1948	Minnesota, USA	Horses, a dog, and wild birds	*Microystis aeruginosa*
1948–1949	Ontario, Canada	Cattle deaths	*Microystis aeruginosa; Anabaena* spp.
1950	Alberta, Canada	A cow, horses, pigs, dogs, turkeys, geese, chickens, and wild birds	*Microcystis*
1949–1951	Manitoba, Canada	Heavy mortality of wild ducks	*Aphanizomenon flos-aquae; Microcystis aeruginosa*
1951	Manitoba, Canada	A horse, 9 dogs	*Aphanizomenon flos-aquae; Microcystis aeruginosa*
1952	Iowa, USA	Thousands of Franklin's gulls, 560 ducks, 400 coots, 200 pheasants, 50 fox squirrels, 18 muskrats, 15 dogs, 4 cats, 2 hogs, 2 hawks, 1 skunk, and 1 mink	*Anabaena flos- aquae*
1953	Lake Semehivichi, USSR	Death of cats, dogs, and waterfowl	*Microcystis aeruginosa*

Table 5. (Continued).

Year	Location	Animals affected	Species producing toxins
1954	Saskatchewan, Canada	Pigs died, cattle unaffected	*Anabaena flos-aquae*
1956	Texas, USA	Fish, frogs, chickens, ducks, turkeys, and cattle died or became ill	*Nostoc rivulae*
1959	Alberta, Canada	14 beef cattle	Blue-green algae
1959	L. Bonney, South Africa	300 sheep, 5 cattle, and 1 horse	Blue-green algae
1959	Saskatchewan, Canada	Approximately 300 dogs, 1 goose, horse, and cattle	*M. aeruginosa/flos-aquae; A. flos-aquae Aphanizomenon*
1961	Saskatchewan, Canada	20 dogs, 3 cattle, perch, and wild ducks	*Anabaena flos-aquae*
1962	Alberta, Canada	1 horse, 8 cows died, 60 cows were sick	Blue-green algae
1962	Saskatchewan, Canada	3 dogs	*Algae*
1963	Rugen, GDR	About 400 ducks	*Nodularia spumigena*
1964	Saskatchewan, Canada	5 dogs	*Anabaena flos-aquae*
1964	Saskatchewan, Canada	20 calves sick, 1 died	*Anabaena, Aphanizomenon, Nodularia*
1964	New Hampshire, USA	Tons of fish died after $CuSO_4$ treatment	*Aphanizomenon flos-aquae*
1965	New South Wales, Australia	20 lambs	*Anacystis cyanea (M. aeruginosa)*

Table 5. (Continued).

Year	Location	Animals affected	Species producing toxins
1965	Saskatchewan, Canada	17 cattle	*Anabaena flos-aquae; Aphanizomenon and Microcystis aeruginosa*
1966	New South Wales, Australia	16 sheep died and 50 were sick	*Anacystis cyanea (M. aerugin-sa)*
1966	Saskatchewan, Canada	2 calves and 1 dog	*Anabaena flos-aquae*
1967	Saskatchewan, Canada	25 pigs	*Anabaena*
1971	New South Wales, Australia	Deaths of honeybees	*Anabaena circinalis*
1972	Alberta, Canada	3 calves	*Anabaena flos-aquae*
1972	Alberta, Canada	12–15 cattle	*Anabaena flos-aquae*
1973–1974	Harbeespoort Dam, South Africa	Cattle deaths	*Microcystis aeruginosa*
1974–1975	South Western and Western Australia	34 sheep, and 52 lambs	*Nodularia spumigena*
1975	New South Wales, Australia	20 lambs	*Anabaena circinalis*
1975	Saskatchewan, Canada	34 cattle	*M. aeruginosa; A. flos-aquae*
1975	Danish coast of the Baltic	30 dogs were sick, 20 died	*Nodularia spumigena*
1976	Washington, USA	4 dogs died, 7 dogs, 1 horse, 1 cow were sick	*Anabaena flos-aquae*

Table 5. (Continued).

Year	Location	Animals affected	Species producing toxins
1977	Montana, USA	8 dogs, and 30 cattle	*Anabaena flos-aquae*
1977	Oklahoma, USA	Several cattle	*Microcystis sp.*
1977	New South Wales, Australia	Turkeys	*Anacystis cyanea*
1978	Rogaland, Norway	4 heifers	*M. aeruginosa*
1978	Cheshire, England	3 cows	*Oscillatoria agardhii*
1979	South Africa	3 rhinoceroses	*Microcystis aeruginosa*
1980	Vaal Dam, South Africa	Cattle	*Microcystis aeruginosa*
1980	New Hampshire, USA	Tadpoles died	*Aphanizomenon flos- aquae*
1981	Illinois, USA	10 sows	*Anabaena spiroides*
1981	New England, Australia	25 sheep died	*Microcystis aeruginosa*
1982	Swedish coast of the Baltic Sea	9 dogs	*Nodularia spumigena*
1983	German coast of the Baltic Sea	16 young cattle	*Nodularia spumigena*
1984	Finnish coast of the Baltic Sea	1 dog and 3 puppies	*Nodularia spumigena*
1984	Goyena, Argentina	72 cows	Hepatotoxins from *Microcystis aeruginosa*

Table 5. (Continued).

Year	Location	Animals affected	Species producing toxins
1984	Montana, USA	11 cattle	*Anabaena flos-aquae; Microcystis aeruginosa; Aphanizomenon flos-aquae*
1984	Queensland, Australia	Death of more than 30 cattle	*Anacystis cyanea*
1985	Alberta, Canada	Approximately 1000 bats, 24 mallards, and American wigeons	Anatoxin-a from *Anabaena flos-aquae*
1985	Wisconsin, USA	9 cows died, 11 were sick	Hepatotoxins from *Microcystis aeruginosa*
1985	South Dakota, USA	9 dogs	Anatoxin-a(s) from *Anabaena flos-aquae*
1985	Aland, Finland	Number of fish, birds, and muskrat deaths	*Oscillatoria agardhii*
1986	Alberta, Canada	16 cows	Anatoxin-a but no indicated species produced anatoxin
1986	Illinois, USA	5 ducks, and 13 pigs died	Anatoxin-a(s) from *Anabaena flos-aquae*
1987	Mississippi, USA	5 cows were sick	*Microcystis aeruginosa*

Table 5. (Continued).

Year	Location	Animals affected	Species producing toxins
1987	Kentucky, USA	20 hogs died	Hepatotoxin from *Anabaena spiroides*
1988	Oklahoma, USA	4 cows, 6 calves, 18 pigs, and 7 ducks died	Hepatoxins and neurotoxin from *M. aeruginosa and A. flos-aquae*
1988	Forez, France	Dead fish	*Microcystis aeruginosa*
1989	Rutland Water, U.K.	Death of 20 sheep, and 14 dogs	Microcystin-LR from *Microcystis aeruginosa*
1989	Saskatchewan, Canada	16 cattle died	Neurotoxin from *Anabaena* sp.
1990	Indiana, USA	2 dogs died	Anatoxin-a from *Anabaena flos-aquae*
1990	Loch Insh, Scotland	3 dogs died	Neurotoxin from *Oscillatoria* sp.
1990	Winhelmhaven, Germany	2 dogs sick and sacrificed	Hepatotoxin from *Nodularia spumigena*
1991	Loch Insh, Scotland	1 dog died	Neurotoxin from *Oscillatoria* sp.
1991	Darling River, Australia	1600 sheep and cattle died	Neurotoxin from *Anabaena circinalis*

Adapted from Ressom et al. (1994).

Table 6. Reported human illness associated with exposure to cyanobacterial toxins.

Study year	Location	Affected population	Exposure route	Sources of toxins	References
1930–1931	Charleston and other townships in West Virginia, USA	9,000 of 60,000 population in Charleston affected by acute gastroenteritis within a 2-wk period	Consuming water from river contaminated by cyanobacteria	Unknown	Veldee 1931; Tisdale 1931
1959	Saskatchewan, Canada	12 people with acute gastroenteritis symptoms	Lakewater contact, swimming	Toxins from toxic *Anabaena*	Dillenberg and Dehnel 1960
1960–1965	Salisbury, Rhodesia (now called Harare, Zimbabwe)	Cases of acute gastroenteritis among children	Consuming water contaminated by cyanobacterial toxins	Unidentified	Zilberg 1966
1975	Allegheny County, Pennsylvania, USA	62% of 8,000 population affected by acute gastroenteritis	Drinking water from open reservoir	Toxins from *Schizothrix calcicola*	Lippy and Erb 1976; Sykora and Keleti, 1981
1979	Palm Island, Queensland, Australia	139 children (2–16 yr) and 10 adults experienced hepatitis-like syndrome with malaise, anorexia, vomiting, tender hepatomegaly, headache, and abdominal pain	Drinking water from open reservoir	Cylindrospermopsin produced by *Cylindrospermopsis raciborskii*	Byth 1980; Bourke et al. 1983; Hawkins et al. 1985
1980–1981	Pennsylvania and Nevada	Over 100 persons affected by skin and eye irritation, earache, hay fever-like symptoms, and acute gastroenteritis	Contact with water from lake (swimming, water skiing)	Toxins from *Anabaena* and *Aphanizomenon*	Carmichael et al. 1985

Table 6. (Continued).

Study year	Location	Affected population	Exposure route	Sources of toxins	References
1983	Armidale region of New South Wales, Australia	Significant increase in the level of serum γ-glutamyltransferase (GGT) in city residents	Consuming water from reservoir contaminated by cyanobacterial blooms	*Microcystis* toxins	Falconer et al. 1983
1989	Staffordshire, UK	2 cases with pneumonia, 16 cases with sore throat, headache, abdominal pain, dry cough, diarrhea, vomiting, and blistered mouths	Reservoir-canoeing with 360° rolls and swimming in water contaminated with *M. aeruginosa*	Toxins from *M. aeruginosa*	Turner et al. 1990; Lawton and Codd 1991
1991	Lake Alexandrina and central Australia	8 cases, 5 adults and 3 children with skin rashes	Water contact	Toxins mainly from *Nodularia* and *Microcystis*	Soong et al. 1992
1992	Outback northern and central Australia	"Barcoo fever" disease characterized by nausea and vomiting at the sight of food	Consuming water contaminated by cyanobacteria	Hepatotoxins produced by cyanobacteria	Hayman 1992
1992	River Murray towns, South Australia	26 cases, aged 1–64 yr with skin, systemic, and multiple symptoms	River water, rainwater tanks, drinking water contact	Toxins from *Anabaena*	El Saadi and Cameron 1993

Table 6. (Continued).

Study year	Location	Affected population	Exposure route	Sources of toxins	References
1994	China	High rate of hepatocellular carcinoma cancer (HCC) in population	Consuming pond/ditch and shallow well water	Microcystins	Yu 1994
1996	Caruaru, Brazil	116 of 130 patients at a dialysis center experienced visual disturbances, nausea, and vomiting; 63 patients died	Receiving dialysis water contaminated by microcystins	Microcystins	Jochimsen et al. 1998

Table 7. Toxins produced by fresh water cyanobacterial and their effects.

Toxins	Molecular structure	Number of variants	Sources	Mode(s) of toxicity
Hepatotoxin				
Microcystins	Cyclic heptapeptide	>50	*Microcystis* *Anabaena* *Nostoc* *Oscillatoria*	Hepatotoxic Tumor promoters Protein phosphatase inhibitors
Nodularin	Cyclic pentapeptide	5	*Nodularia*	Hepatotoxic Tumor promoters Protein phosphatase inhibitors
Cylindrospermopsin	Alkaloid		*Cylindrospermopsis raceborskii* *Umezkia natans*	Cytotoxic Liver and organ damage
Neurotoxin				
Anatoxin-a	Secondary amine alkaloid	1	*Anabaena*	Neurotoxic, depolarizing neuro-muscular blocker
Anatoxin-a(s)	Guanidinium methyl phosphate ester	1	*Anabaena*	Neurotoxic Cholinesterase inhibitor
Saxitoxin	Alkaloid	~6	*Aphanizomenon*	Neurotoxic, sodium channel blockers
Lipopolysaccharide (LPS) endotoxins	Lipopolysaccharides	>3	*Microcystis* *Oscillatoria*	Toxic shock Gastroenteritis Inflammation

Adapted from Hawkins et al. (1985); Ohtani and Moore (1992); Harada et al. (1994a); Codd (1995).

sively studied and the data for microcystins are probably sufficient. Data related to other toxins are comparatively limited.

A. Dose–Response Relationships of Microcystins

Microcystins have attracted concern from public health authorities because of their health effects, wide occurrence, and persistence in the environment. Microcystins have been tested in a variety of animal species through ip, oral, dermal, and intranasal routes. Like other toxins, microcystins appear to be more toxic when test animals are administered through the ip route (ip LD_{50} is approximately 25–250 µg/kg in mice, compared to an oral LD_{50} of approximately 5 mg/ kg mice) (Fawell et al. 1993). Most studies have shown that increasing doses of microcystins can lead to increase in liver damage, characterized by elevation of some liver enzymes, hepatocyte damage, and necrosis. Yoshida et al. (1997) have studied the toxic effects of purified microcystin-LR (>95% purity) on groups of five female mice that were intraperitoneally injected with 40–82.9 µg/kg bw. The authors found that mortality rate increased as a result of increased exposure. By using the Spearman–Käber method, LD_{50} of purified microcystin-LR in this experiment was estimated at 65.46 µg/kg bw mice (with 95% confident limits of 54.48–73.28). The mortality and dose–response curve are illustrated in Fig. 9 (see page 37).

Even though exposure through the oral route is less toxic than that of the ip route, the oral route has attracted greater concern because it is the major route by which humans may be exposed to microcystins and other cyanobacterial toxins through drinking water. Three studies have been reported on chronic exposure to microcystins. One study was conducted in England (Fawell et al. 1993) and the others in Australia (Falconer et al. 1981, 1994). Falconer et al. (1981) reported a

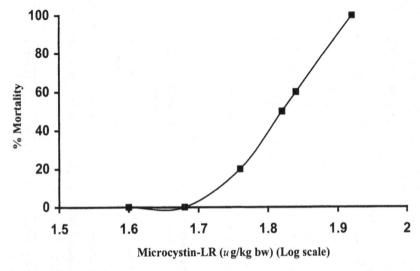

Fig. 9. Dose–response curve for female mice after ip exposure to purified microcystin-LR.

study of microcystins orally administered to mice for up to 1 yr in which increasing exposure dose resulted in increased mortality rate (particularly in male mice) and chronic liver injury. The cumulative mortality of male mice exposed to microcystins for up to 37 wk (Fig. 10) shows that the mortality rate clearly increases with increasing doses. Mortality increases rapidly when the dose is higher than 0.5 mgkg^{-1}d^{-1}.

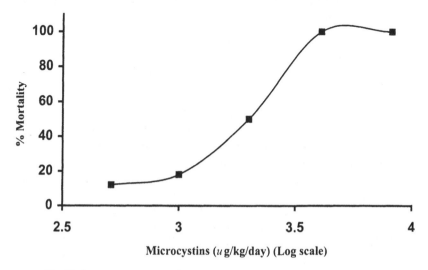

Fig. 10. Dose–response curve of microcystins for mice after 37-wk oral exposure.

In a study by Fawell et al. (1993), groups of 15 male and female mice were orally exposed to 0, 40, 200, and 1000 µg microcystin-LR/kg bw/d. After a 13-wk exposure, no mortality caused by the treatment was recorded. No significant changes were observed in mice exposed to 40 µg microcystin-LR/kg bw/d. Therefore, this dose is considered as the oral no-observable-adverse-effect level (NOAEL) of microcystin-LR (Fawell et al. 1993). At higher doses, a number of significant changes in blood chemistry occurred and a reduction in body weight gain was reported. Even though Fawell et al. (1993) reported that there was no dose-related response, these changes in blood chemistry clearly resulted from microcystin-LR treatment.

Falconer et al. (1994) also found changes in plasma samples collected from pigs orally exposed to *M. aeruginosa* extracts. In this experiment, groups of five pigs were orally administered with the *Microcystis* extract, equivalent to 280–1312 µg total microcystins/kg bw/d (toxin content of the extracts was estimated based on ip LD$_{50}$ in mice). Toxicity of the extracts was also determined by high performance liquid chromatography (HPLC) and protein phosphatase inhibition assay (Falconer et al. 1994). The equivalent concentrations given to test pigs were found to range from 184 to 860 µg/kg bw/d (if the toxin content of extract was determined by HPLC) or 88 to 441 μg/kg bw/d (if the toxin content of the extract was based on protein phosphatase inhibition assay). The lowest dose was reported as the NOAEL (Falconer et al. 1994). Because tissue damage was observed in one pig (1 of 5), however, this dose will be considered as the lowest-observable-adverse-effect level (LOAEL) in this review and will be used to calculate the guideline values for microcystins in drinking water. At higher doses, plasma indicators that showed changes with increasing exposure to *Microcystis* toxin were GGT (γ-glutamyl transpeptidase), ALP (alkaline phosphatase), total bilirubin, and plasma albumin.

B. Dose–Response Relationships of Cylindrospermopsin

Although isolated later than other known cyanobacterial toxins, cylindrospermopsin has also attracted the concern of public health authorities in Australia and other parts of the world. Cylindrospermopsin is currently believed to occur in tropical regions and has been studied in Australia, especially in Queensland. Currently, no study has been done on oral subchronic or chronic exposure to this toxin. Available data mainly concern acute and subacute exposure to the toxin through ip and oral route. In a study conducted on mice, it was found that mice intraperitoneally challenged with approximately 0.02 mg/kg bw/d for 12 d showed no pathological changes in the liver, kidney, heart, or lungs (Shaw et al., unpublished data). Therefore, this dose will be considered as the NOAEL of the ip route. By using the Spearman–Käber method (Wardlaw 1985), the ip 7-d LD_{50} of cylindrospermopsin was estimated to be 0.17 mg/kg mice (with 95% confident limits of 0.1–0.28). This value falls within the range of the ip LD_{50} of microcystins, suggesting that the toxicity of cylindrospermopsin is probably as potent as microcystins. The study also showed that increasing injury in liver corresponded to increasing exposure dose. In terms of cumulative mortality, 100% death was observed in mice orally administered with 8 mg cylindrospermopsin/kg, but no mortality was recorded at lower doses. By ip administration, however, deaths were observed in mice injected with doses greater than 0.05 mg/kg mice (Fig. 11).

VIII. Exposure Assessment for Freshwater Cyanobacterial Toxins

A. Occurrence of Freshwater Cyanobacterial Toxins

1. *Occurrence of Dissolved Toxins in Water Contaminated by Cyanobacterial Blooms.* The occurrence of cyanobacterial toxins in the freshwater environment is unpredictable and has been shown to vary temporally and spatially (Ueno et al. 1996). In general, the highest toxin concentrations in the water have been found during summer and early autumn. Among cyanobacterial toxins, only microcystins appear to occur extensively around the world. The occurrence of cylindrospermopsin has been increasingly reported in Australia, particularly in Queensland. Recently, cylindrospermopsin has been detected in Sri Lanka and probably occurs in other tropical and temperate regions as well. Neurotoxins have been

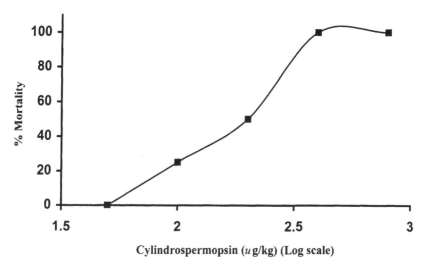

Fig. 11. Dose–response curve for mice after ip exposure to cylindrospermopsin.

detected in water bodies in Denmark and Australia. There are few reports of the occurrence of other toxins.

Microcystins have been detected in a range of water bodies, including pond or ditch water and lake, river, and reservoir water as well as shallow well water. The concentrations of dissolved microcystins in water ranged from trace to a maximum of 55 μg/L detected by an enzyme-linked immunosorbent assay (ELISA) method in lake water of Dingshan in Shanghai, China (Yu 1994). Data obtained from the Queensland Department of Natural Resources and the South East Queensland Water Board showed that the level of cylindrospermopsin ranged from 0.2 to 10.1 μg/L in storages in Queensland during 1997–1998 (McGregor and Kinros, unpublished data).

2. *Occurrence of Total Cyanobacterial Toxins in Raw Water Contaminated with Blooms After Algicide Treatment.* It is generally agreed that application of algicide to treat cyanobacterial blooms lyse cyanobacterial cells and so release toxins into water. Most studies have shown that a large proportion of microcystins is contained in cells, and so the toxin concentration in water would be much higher if algicide was applied. Jones and Orr (1994) studied the toxins released from *M. aeruginosa* in an Australian recreational lake and found that the toxin concentration in the lake after algicide treatment was much higher (approximately 210–310 times) than the toxin levels detected in water before treatment.

3. *Potential for the Presence of Cyanobacterial Toxins in Treated Water.*
Background Information of Current Water Treatment Processes. Currently, a variety of processes have been applied in water treatment (Keijola et al. 1988; Falconer et al. 1989; Himberg et al. 1989; Drikas 1994; Lambert 1993; Nicholson et al. 1994). For example:

1. Conventional treatment process (chemical coagulation–sand filtration–chlorination)
2. Prechlorination–pH adjustment–flocculation with alum–gravity sedimentation–rapid sand filtration–postchlorination, fluoridation, and reticulation
3. $Al_2(SO_4)_3$ flocculation–sand filtration–chlorination
4. $FeCl_3$ flocculation–sand filtration–chlorination
5. Addition of powdered activated carbon (PAC)-$Al_2(SO_4)_3$ flocculation–sand filtration–chlorination
6. $Al_2(SO_4)_3$ flocculation–sand filtration–granular activated carbon (GAC) filtration–chlorination
7. Oxidation (chlorination, chloramination, ozonation)-$Al_2(SO_4)_3$ flocculation–sand filtration–chlorination

The Efficiency of Current Water Treatment Processes in Removing Cyanobacterial Toxins. According to laboratory- and pilot-scale studies as well as current practices, conventional treatment processes appear to be ineffective in removing cyanobacterial toxins. Some studies claimed that the conventional treatment process (coagulation/flocculation–sand filtration–chlorination) could remove some cyanobacterial toxins (Keijola et al. 1988; Himberg et al. 1989). However, the efficiency was too low, 11%–18% (Keijola et al. 1988; Himberg et al. 1989). Therefore, with the mass occurrence of toxins in water, the conventional treatment process would fail to protect water consumers from exposure to high concentration of cyanobacterial toxins through drinking water.

Poor efficiency of removal of cyanobacterial toxins (17%) was also obtained when the chloramination was applied with a concentration of 20 mg/L and a contact time (Ct) of 5 d (Nicholson et al. 1994). Similar efficiency (11%–18%) was reported when 0.5 mg/L NaOCl with Ct of 20 min and pH of 5.5–6.6 were used (Keijola et al. 1988; Himberg et al. 1989). No removal of microcystins was observed with an application of 5 mg/L $Ca(OCl)_2$ with Ct of 30 min and pH of 8.5 (Hoffman 1976). In contrast, Nicholson et al. (1994) found that the efficiency of microcystin removal was more than 95% when they used Cl_2 and $Ca(OCl)_2$, both with a concentration of 1 mg /L and Ct of 30 min. However, the proportion of toxins removed was 70%–80% when the authors used 5 mg/L NaOCl and Ct of 30 min (Nicholson et al. 1994). The process of prechlorination (1 mg/L)–flocculation with alum–sedimentation–sand filtration–chloramination has been demonstrated to remove cylindrospermopsin to below the detection limit (<0.2 μg/L) of HPLC-mass spectrometry (HPLC-MS/MS), and the contact time is approximately 1 hr (R Gray; Brisbane Water, personal communication 1997).

Ozonation (1 mg ozone/L, Ct=30 min + sand filtration + Cl_2) was shown to be effective in removing cyanobacterial toxins (99% or higher) (Keijola et al. 1988; Himberg et al. 1989; Drikas 1994). The efficiency of this treatment process depends on the concentration of ozone used (Drikas 1994), and this process has not been widely applied. Activated carbon has been shown to be effective in removing cyanobacterial toxins and granular activated carbon (GAC) appeared to be more effective than powdered activated carbon (PAC) (Himberg et al. 1989; Falconer et al. 1989; Drikas 1994). Up to 100% toxin removal efficiency may be achieved by this process. The efficiency also depends on the type of activated carbon used and appears to be reduced with continued use. Activated carbon is expensive and thus is not used widely. In Australia, for example, only small water treatment plants appear to use activated carbon.

The Occurrence of Cyanobacterial Toxins in Treated Water. Despite the fact that some of the water treatment processes are believed to be able to remove cyanobacterial toxins to levels that pose insignificant risk to water consumers, cyanobacterial toxins are occasionally detected in treated water (Table 8). Furthermore, the potential toxicities of transformation products and by-products of these treatments are still largely unknown.

B. Exposure Scenarios for Humans

1. *Exposure to Freshwater Cyanobacterial Toxins Through Raw Water Contaminated by Toxic Blooms.*

Exposure Through Drinking Raw Water. Because raw water contaminated by cyanobacteria may contain high concentration of toxins, drinking this water may pose risks to consumers. This situation can occur in areas where water treatment processes are not adequate or efficient such as in less developed countries, particularly in rural areas. In these areas, a high proportion of the population may consume surface water without treatment. When cyanobacterial toxins occur, these populations may experience both acute and chronic exposure to the toxins. The consequence of chronic exposure to cyanobacterial toxins has been demonstrated in China where a high percentage of patients suffered from primary liver cancer that is believed to be associated with consumption of pond and ditch water contaminated by microcystins (Yu 1989, 1995). Drinking untreated raw water may transfer both dissolved toxins and toxins contained in cells into the bodies of the consumers, resulting in elevated toxin exposure.

Exposure Through Raw Water Used for Other Domestic Purposes. Cyanobacterial toxins can enter human bodies through pathways other than the oral route, including inhalation during showering or water contact. Illness in humans associated with inhaling microcystins had been reported by Lawton and Codd (1991). The intranasal route appeared to be as toxic as the intraperitoneal route (Fitzgeorge et al. 1994). Therefore, the risk posed by inhaling cyanobacterial toxins during showering should be of concern. Apart from showering, other activities such as washing and other domestic operations result in contact with water containing free toxins or toxic cyanobacterial cells that pose risks to human health.

Exposure Through Hemodialysis. A recent cyanobacterial poisoning through dialysate water in Caruaru, Brazil, has suggested that hemodialysis is another possible route by which humans can be exposed to cyanobacterial toxins. The occurrence of cyanobacterial toxins in dialysate water may lead to mortality in patients who are undergoing hemodialysis treatment.

Exposure Through Recreational Activities. Recreational activities in water contaminated by cyanobacterial blooms and their toxins may pose a risk through the oral route, dermal route, and intranasal route. During recreational activities, people, particularly children, may accidentally ingest cyanobacterial scum or water

Table 8. Occurrence of cyanobacterial toxins in treated water.

Toxins	Concentration (µg/L)	Detection methods	Sources	References
Microcystins	0.001–0.023	Enzyme-linked immunosorbent assay	Treated water from clear lake water, sampled in 1993	Falconer 1993a
Microcystin-LR	0.1–1	Liquid chromatography-linked protein phosphatase inhibition assay	Typical levels detected in samples of tap water supplied by Little Beaver Lake, Alberta, Canada, during 08/92	Holmes et al. 1994
Microcystins	0.05–0.12	Protein phosphatase inhibition assay	Tested drinking water sampled on 26/08–02/10/1992 from the Ferintosh treatment plant, Canada, applying full-scale treatment (coagulation-sedimentation, dual media filtration, and chlorination combined with granular activated carbon filtration)	Lambert et al. 1994
Microcystins	0.09–0.18	Protein phosphatase inhibition assay	Tested drinking water sampled on 26/08–02/10/1992 from the Camrose treatment plant, Canada, applying full-scale treatment (coagulation-sedimentation, dual media filtration and chlorination combined with powdered activated carbon filtration)	Lambert et al. 1994
Microcystins	0.1–0.5	High performance liquid chromatography	Detected in drinking water in the Taalintehdas community in 1989	Lahti 1997
Microcystins	0.3–0.4	High performance liquid chromatography	Detected in sand-filtered water when the density of $Microcystis$ cells was $1–3 \times 10^5$ cells/L	Lahti 1997

containing free toxins. Some activities such as canoeing, windsurfing, and swimming can transfer toxins through inhalation. Even though cyanobacterial toxins are believed to be unable to cross the skin, the study of Wannemacher et al. (1987) showed that microcystin-LR was toxic to guinea pigs via the dermal route. Furthermore, some cyanobacterial species are able to produce LPS endotoxins that may cause eye, skin irritation, and gastroenteritis. Therefore, recreational activities represent another vehicle that may transfer cyanobacterial toxins to humans.

2. *Exposure to Freshwater Cyanobacterial Toxins Through Treated Water.* The findings from laboratory- and pilot-scale studies show that the concentration of cyanobacterial toxins can be reduced by water treatment processes, and so water consumers may be exposed to lower toxin concentrations compared to those contained in raw water. Nevertheless, the efficiency in removing toxins varies with types of treatment (see earlier). Moreover, by-products or transformation products of treatment processes are still a potential cause for concern.

Exposure Through Drinking Water Treated by Conventional Treatment Processes. These conventional processes appear to be ineffective in removing cyanobacterial toxins, although they can remove effectively intact algal cells. Therefore, consumers may still be exposed to dissolved toxins that occur in the raw intake water. During mass occurrence of cyanobacterial blooms such as in summer or autumn, conventional treatment processes would not be able to protect water consumers from high exposure to cyanobacterial toxins.

Exposure Through Drinking Water Treated with Activated Carbon. Activated carbon has been shown to be effective in removing cyanobacterial toxins. Nonetheless, the efficiency depends on the type of activated carbon employed and the period of use. Studies in two water treatment plants using PAC and GAC in Canada showed that water after being treated still contained 0.09–6.4 μg of microcystins per liter (Hrudney et al. 1994b). The efficiency of the plants was only 70%–80%, indicating that consumers of water treated by activated carbon may still be exposed to cyanobacterial toxins.

Exposure Through Drinking Water Treated with Oxidants. The main oxidants that are currently studied for use in removing cyanobacterial toxins include ozone, chlorine, potassium permanganate, and hydrogen peroxide. Of these, ozone appears to be most effective in removing cyanobacterial toxins. However, the efficiency appears to vary with dosage, and the risks associated with by-products have not been properly assessed. Chlorine at a concentration of 1 mg/L or more has the potential to reduce microcystins in water (Nicholson et al. 1994). Nevertheless, at this concentration, trihalomethane and other organic chlorine compounds may be formed that may themselves have detrimental effects.

3. *Exposure Through Food Consumption.* Food consumption is another vehicle that may transfer toxins to humans. Cyanobacterial toxins may be accumulated in some aquatic organisms such as mussels (Vasconcelos 1995). In turn, these food organisms may cause poisoning in consumers. Information on this particular aspect, which is generally scarce, is urgently needed for risk assessment purposes.

IX. Risk Characterization for Humans

The purpose of this section is to calculate the possible guideline values for cyano-bacterial toxins and to predict which populations are at risk from these toxins. The findings of this section may be considered in the context of risk communication and risk management so that risk posed by the toxins can be mitigated. On the basis of data available, only guideline values of anatoxin-a, microcystins, microcystin-LR, and cylindrospermopsin can be established. The reason to derive guideline values for both microcystins and microcystin-LR is that microcystin-LR is the most toxic microcystin known to date and its tumor-promoting property has been confirmed. Therefore, the guideline values of microcystin-LR can be considered as the lowest limit concentration of microcystins in drinking water and the guideline also takes tumor promotion into account. The guideline values for microcystins are based only on toxicity because some microcystins may not possess tumor-promoting properties.

A. Derivation of Possible Guideline Values

1. *Sources and Evaluation of Data.* Currently, there are few studies on chronic or subchronic exposure to neurotoxins. Only Astrachan et al. (1980) have conducted subacute studies of anatoxin-a on mice. In this study, mice orally administered with 0.51 mg anatoxin-a/kg/d for 7 wk showed no apparent effects, and this dose is used in this study as the NOAEL to calculate the guideline values for anatoxin-a.

For microcystins, three subchronic studies have been conducted on mice and pigs. The first study was carried out on mice for up to a year (Falconer et al. 1988). The dose of 0.5 mg/kg bw/d was used to calculate a possible guideline value (Carmichael and Falconer 1993). Nevertheless, this dose is not reliable enough to establish the guideline values because mortality was observed at this dose. The second study was conducted by Fawell et al. (1993), and an oral NOAEL of 40 µg/kg bw/d was obtained after 13 wk of oral administration with microcystin-LR on mice. Because microcystin-LR is the most toxic of all micro-cystins investigated so far and its tumor-promoting property has been confirmed, this NOAEL will be used to set up the guideline value that takes tumor promotion into account. The last study was carried out on pigs exposed to *Microcystis* extracts for 44 d (Falconer et al. 1994). In this study, groups of five pigs were administered with 184, 522, and 860 µg microcystins/kg bw/d for 44 d. The toxin contents of *Microcystis* extract used were determined by HPLC analysis. At 184 $\mu g\ kg^{-1}d^{-1}$, one pig showed tissue damage and the other appeared normal. There-fore, this dose is used as LOAEL to derive the guideline value for microcystins based on toxicity.

For cylindrospermopsins, because no oral subchronic study has been reported, data obtained from intraperitoneal administration with subchronic expo-sure (12 d) are used to calculate a possible guideline value (Shaw et al., unpub-lished data). In this study, mice injected with 0.02 mg cylindrospermopsin per kg/d appeared normal, and so this dose is applied as the NOAEL.

2. *Methods.* Possible guideline values are calculated from the LOAEL or NOAEL as follows:

$$TDI = \frac{NOAEL \text{ or } LOAEL}{UF}$$

$$GV = \frac{TDI \times bw \times P}{C}$$

Where TDI = tolerable daily intake (mg/kg bw/d or µg/kg bw/d)

Where NOAEL = no-observable-adverse-effect level (mg/kg bw/d or µg/kg bw/d)

Where LOAEL = lowest-observable-adverse-effect level (mg/kg bw/d or µg/kg bw/d)

Where bw = body weight, assuming that the average weights of adults, children, and infants are 60, 10, and 5 kg, respectively

Where C = daily drinking water consumption; assuming 2 L for a 60-kg adult, 1 L for a 10-kg child, and 0.75 L for a 5-kg infant

Where P = percentage of the TDI allocated to drinking water; for cyanobacterial toxins, because there is evidence of bioaccumulation of the food chain for some toxins, exposure through food consumption may be possible, and therefore P value in this case is assumed to be 80% (0.8)

Where UF = uncertainty factor

Where GV = guideline value

To establish the guideline for anatoxin-a, UF of 1000 is used (10 for interspecies variation, 10 for intraspecies variation, and 10 for less-than-lifetime study). In the case of microcystins, UF of 5000 is used (10 for interspecies variation, 10 for intraspecies variation, 5 for using LOAEL, and 10 for less-than-lifetime study). A UF of 3000 is used to calculate the guideline value for microcystin-LR that takes tumor promotion into account (10 for interspecies variation, 10 for intraspecies variation, 10 for less-than-lifetime study, and 3 for accommodating tumor promotion). For cylindrospermopsin, UF of 1000 is used (10 for interspecies variation, 10 for intraspecies variation, and 10 for less-than-lifetime study).

3. *Calculation of Guideline Values (GV).*

Guideline Values for Anatoxin-A.

$$TDI = \frac{NOAEL}{UF} = \frac{0.51(mg/kg/d)}{1000} = 0.51 \mu g/kg/d$$

For adults:

$$GV_{ad} = \frac{0.51 \times 60 \times 0.8}{2} = 12.24 \mu g/L$$

For children:

$$GV_{chd} = \frac{0.51 \times 10 \times 0.8}{1} = 4.08 \mu g/L$$

For infants:

$$GV_{inf} = \frac{0.51 \times 5 \times 0.8}{0.75} = 2.72\,\mu g/L$$

Guideline Values for Microcystins.

$$TDI = \frac{LOAEL}{UF} = \frac{184(\mu g/kg/d)}{5000} = 0.0368\,\mu g/kg/d$$

For adults:

$$GV_{ad} = \frac{0.0368 \times 60 \times 0.8}{2} = 0.88\,\mu g/L$$

For children:

$$GV_{chd} = \frac{0.0368 \times 10 \times 0.8}{1} = 0.29\,\mu g/L$$

For infants:

$$GV_{inf} = \frac{0.0368 \times 5 \times 0.8}{0.75} = 0.20\,\mu g/L$$

Guideline Values for Microcystin-LR.

$$TDI = \frac{NOAEL}{UF} = \frac{40(\mu g/kg/d)}{3000} = 0.0133\,\mu g/kg/d$$

For adults:

$$GV_{ad} = \frac{0.0133 \times 60 \times 0.8}{2} = 0.32\,\mu g/L$$

For children:

$$GV_{chd} = \frac{0.0133 \times 10 \times 0.8}{1} = 0.11\,\mu g/L$$

For infants:

$$GV_{inf} = \frac{0.0133 \times 5 \times 0.8}{0.75} = 0.07\,\mu g/L$$

Guideline Values for Cylindrospermopsin.

$$TDI = \frac{NOAEL}{UF} = \frac{0.02(mg/kg/d)}{1000} = 0.02\,\mu g/kg/d$$

For adults:

$$GV_{ad} = \frac{0.02 \times 60 \times 0.8}{2} = 0.48\,\mu g/L$$

For children:

$$GV_{chd} = \frac{0.02 \times 10 \times 0.8}{1} = 0.16 \mu g/L$$

For infants:

$$GV_{inf} = \frac{0.02 \times 5 \times 0.8}{0.75} = 0.11 \mu g/L$$

The tolerable daily intake and possible guideline values of various toxins are summarized in Table 9.

B. Significance of the Guideline Values

These guideline values suggest that humans are more susceptible to microcystins than anatoxin-a. However, it is not possible to compare microcystins and cylindrospermopsin because guideline values of cylindrospermopsin are based on ip data. As mentioned, however, toxicity of cylindrospermopsin is comparable to that of microcystins. On this basis, it may be conceivable that exposure to the same amount of cylindrospermopsin and microcystins may lead to similar consequences.

Guideline values for microcystin-LR may possibly be used in the monitoring of nodularin levels in water because of their similar toxicity. Nevertheless, safe levels of nodularin in water may be lower than that of microcystin-LR because nodularin is suspected to be a new environmental carcinogen (Kaya 1996).

The possible guideline values for microcystins and microcystin-LR are comparable to the guideline value (for adults) that has recently been developed by the World Health Organization (WHO) (1 μg/L for microcystins based on toxicity and 0.3 μg/L if tumor promotion is taken into account). The guideline value of 0.88 μg microcystin/L for adults is also similar to the value calculated by Falconer et al. (1994) using the same data in their pig experiment. The value of 0.32 μg of microcystin-LR/L is lower than the value of 0.5 μg/L obtained by Kuiper-Goodman et al. (1994), even though two guideline values were set up based on the same NOAEL. This difference occurs because Kuiper-Goodman et al. (1994) did not consider the possible consumption of the toxin from other routes and so 100% of the TDI was allocated to drinking water. Because microcystins may accumulate in aquatic organisms, the consumption of these contaminated organisms may transfer the toxins into the human body. Therefore, exposure through routes other than drinking water should be considered. Hence, the percentage of TDI allocated to drinking water should be lower than 100%.

Based on the guideline values in Table 9, it is apparent that populations consuming surface water without treatment may be exposed to high concentrations of cyanobacterial toxins during the peak season and thus the risk posed by the toxins to these populations is quite high, compared with populations consuming treated water. On this basis, risks posed by cyanobacterial toxins to populations in less developed countries, particularly in rural areas, may be higher than those liv-

T. N. Duy et al.

Table 9. Tolerable daily intake (TDI) and guideline value (GV) for anatoxin-a, microcystins, microcystin-LR, and cylindrospermosins in drinking water.

TDI and GV for anatoxin-a in drinking water:

Groups	TDI (μg/kg/d)	GV (μg/L)
Infants (5 kg bw)	0.51	2.72
Children (10 kg bw)	0.51	4.08
Adults (60 kg bw)	0.51	12.24

TDI and GV for microcystins in drinking water:

Groups	TDI (μg/kg/d)	GV (μg/L)	GV (cells/mL)[a]
Infants (5 kg bw)	0.0368	0.20	1000
Children (10 kg bw)	0.0368	0.29	1450
Adults (60 kg bw)	0.0368	0.88	4400

TDI and GV for microcystin-LR in drinking water (guideline values take both toxicity and tumor promotion into account):

Groups	TDI (μg/kg/d)	GV (μg/L)	GV (cells/mL)[b]
Infants (5 kg bw)	0.0133	0.07	350
Children (10 kg bw)	0.0133	0.11	500
Adults (60 kg bw)	0.0133	0.32	1600

TDI and GV for cylindrospermopsin in drinking water:

Groups	TDI (μg/kg/d)	GV (μg/L)	GV (cells/mL)[c]
Infants (5 kg bw)	0.02	0.11	4231
Children (10 kg bw)	0.02	0.16	6154
Adults (60 kg bw)	0.02	0.48	18461

[a]Assuming 1 cyanobacterial cell contains 0.2 pg of toxins (Falconer et al. 1994).
[b]Assuming 1 cyanobacterial cell contains 0.2 pg of toxins (Falconer et al. 1994).
[c]Assuming 1 cyanobacterial cell contains 0.026 pg of toxins (Hawkins et al. 1997).

ing in developed countries. Because the toxicity of the by-products or transformation products of the toxin treatment processes are still unknown, however, populations consuming treated water are still at risk. Children and infants are the most susceptible groups and at peak levels of cyanobacterial toxins, even treated municipal water supplies may pose risks to them.

Table 9 also shows the guideline values for cell numbers. However, the guideline values based on cell numbers are less reliable because the toxicity of cyanobacterial blooms varies considerably. High toxin concentration may occur in the early phase of a bloom, while certain cells in a severe bloom may not be toxic. Therefore, best practice should consider both cell numbers and toxin concentration.

X. Risk Characterization of Freshwater Cyanobacterial Toxins for Domestic Animals

Unlike human beings, both wild and domestic animals, including aquatic animals, can be exposed to high concentrations of cyanobacterial toxins. First of all, aquatic animals, because of their habitats, can come into direct contact with the toxins that are released from the cells. Therefore, they may be exposed to the highest concentrations before the toxins begin to degrade. Likewise, wild and domestic animals may be exposed to the toxins through water containing free toxins or through ingesting scum containing toxins. Because of the limitation of the data, risk characterization is only carried out for some domestic animals.

A. Derivation of Possible Guideline Values for Some Domestic Animal Groups

1. *Sources of Data and Methods.* Sources of data and methods are similar to those employed for humans in Section IX. However, the UF used to set up guideline values for microcystins is 500 (10 for intraspecies variation, 10 for less-than-lifetime study, and 5 for using LOAEL). UF of 300 (10 for intraspecies variation, 10 for less-than-lifetime study, and 3 for accommodating tumor promotion) is used to calculate the guideline value for microcystin-LR; UF of 100 is used for anatoxin-a and for cylindrospermopsin. Estimates of body weights and water consumption rates are summarized in Table 10.
2. *Calculation of Guideline Values.* The guideline values of anatoxin-a, microcystins, microcystin-LR, and cylindrospermopsin for some domestic animals are presented in Tables 11–14.

B. Significance of Guideline Values

From the guideline values, it is clear that livestock may be at high risk when toxic blooms occur. Animals often have access to the water along the shoreline where blooms tend to accumulate. Based on cell count guideline values, microcystins appear to pose higher risk to livestock compared with anatoxin-a and cylindrospermopsins. Among the livestock groups examined, calves seem to be the most susceptible to the cyanobacterial toxins because they have lower body weight but water consumption in this group is quite high. In contrast, lambs fattening on irrigated pasture seem to be the most tolerant to the toxins because of their low water consumption. The guideline values show that microcystin-LR is the most potent freshwater cyanobacterial toxin that is likely to pose a risk to livestock. Because of their daily consumption of a large amount of water, these livestock animals may be exposed to high toxin levels. Attention should be paid when bloom occurs, even at relatively low cell densities. Because of the lack of information related to other cyanobacterial toxins, the guideline values are incomplete. However, data from laboratory and field studies as well as animal poison-

Table 10. Body weights and water consumption by some livestock.

Animals	Body weight (kg)	Water consumption (L/d)
Cattle		
Dairy cows in milk	500	70
Dairy cow dry	500	45
Beef cattle	450	45
Calves	60	22
Horses		
Working horse	600	55
Grazing horse	600	35
Sheep		
Nursing ewes on dry feed	50	9
Mature sheep on dry pasture	50	7
Mature sheep on irrigated pasture	50	3.5
Fattening lambs on dry pasture	50	2.2
Fattening lambs on irrigated pasture	50	1.1
Pigs		
Brood sows	160–200	22
Mature pigs	160–200	11
Poultry		
Laying hens	3	0.32
Nonlaying hens	3	0.18
Turkey	5–8	0.55

Source: Hasker P, Animal Research Institute, Department of Primary Industry, Queensland Australia, personal communication.

Table 11. Tolerable daily intake (TDI) and guideline value (GV) of anatoxin-a in drinking water for some domestic animal groups.

Animals	TDI	GV (µg/L)
Cattle		
Dairy cows in milk	5.1	36
Dairy cow dry	5.1	57
Beef cattle	5.1	51
Calves	5.1	14
Horses		
Working horse	5.1	56
Grazing horse	5.1	88
Sheep		
Nursing ewes on dry feed	5.1	28
Mature sheep on dry pasture	5.1	36
Mature sheep on irrigated pasture	5.1	73
Fattening lambs on dry pasture	5.1	116
Fattening lambs on irrigated pasture	5.1	232

Table 11. (Continued).

Animals	TDI	GV (μg/L)
Pigs		
Brood sows	5.1	37–46
Mature pigs	5.1	74–93
Poultry		
Laying hens	5.1	48
Nonlaying hens	5.1	85
Turkey	5.1	46–74

Assuming percent of the TDI allocated to drinking water is 100% (P=1).

Table 12. Tolerable daily intake (TDI) and guideline value (GV) of microcystins in drinking water for some domestic animal groups.

Animals	TDI	GV (μg/L)	GV (cells/mL)
Cattle			
Dairy cows in milk	0.368	3	13,150
Dairy cow dry	0.368	4	20,450
Beef cattle	0.368	4	18,400
Calves	0.368	1	5,000
Horses			
Working horse	0.368	4	20,050
Grazing horse	0.368	6	31,550
Sheep			
Nursing ewes on dry feed	0.368	2	10,200
Mature sheep on dry pasture	0.368	3	13,150
Mature sheep on irrigated pasture	0.368	5	26,300
Fattening lambs on dry pasture	0.368	8	41,800
Fattening lambs on irrigated pasture	0.368	17	83,600
Pigs			
Brood sows	0.368	2.7–3.4	13,400–16,750
Mature pigs	0.368	5.4–6.7	26,750–33,450
Poultry			
Laying hens	0.368	3	17,250
Nonlaying hens	0.368	6	30,650
Turkey	0.368	3–5	16,750–26,780

Guideline values do not take tumor promotion into account.
Assuming 1 cell contains 0.2 pg toxins (Falconer et al. 1994).
Percent of the TDI allocated to drinking water is 100% (P=1).

Table 13. Tolerable daily intake (TDI) and guideline value (GV) of microcystin-LR in drinking water for some domestic animal groups.

Animals	TDI	GV (µg/L)	GV (cells/mL)
Cattle			
Dairy cows in milk	0.133	0.9	4,750
Dairy cow dry	0.133	1.5	7,400
Beef cattle	0.133	1.3	6,650
Calves	0.133	0.4	1,800
Horses			
Working horse	0.133	1.5	7,250
Grazing horse	0.133	2.3	11,450
Sheep			
Nursing ewes on dry feed	0.133	0.7	3,700
Mature sheep on dry pasture	0.133	1.0	4,750
Mature sheep on irrigated pasture	0.133	1.9	9,500
Fattening lambs on dry pasture	0.133	3.0	15,150
Fattening lambs on irrigated pasture	0.133	6.1	30,303
Pigs			
Brood sows	0.133	1.0–1.2	4,850–6,050
Mature pigs	0.133	1.9–2.4	9,700–12,100
Poultry			
Laying hens	0.133	1.3	6,250
Nonlaying hens	0.133	2.2	11,100
Turkey	0.122	1.2–1.9	6,050–9,700

Guideline values take tumor promotion into account.
Assuming 1 cell contains 0.2 pg toxins (Falconer et al. 1994).
Percent of the TDI allocated to drinking water is 100% (P=1).

Table 14. Tolerable daily intake (TDI) and guideline value (GV) of cylindrospermopsin in drinking water for some domestic animal groups.

Animals	TDI	GV (µg/L)	GV (cells/mL)
Cattle			
Dairy cows in milk	0.2	1.4	55,000
Dairy cow dry	0.2	2.2	85,385
Beef cattle	0.2	2.0	76,923
Calves	0.2	0.6	21,154
Horses			
Working horse	0.2	2.2	83,846
Grazing horse	0.2	3.4	131,923

Table 14. (Continued).

Animals	TDI	GV (µg/L)	GV (cells/mL)
	(cont.)		
Sheep			
Nursing ewes on dry feed	0.2	1.1	42,692
Mature sheep on dry pasture	0.2	1.4	55,000
Mature sheep on irrigated pasture	0.2	2.9	110,000
Fattening lambs on dry pasture	0.2	4.6	175,000
Fattening lambs on irrigated pasture	0.2	9.1	349,615
Pigs			
Brood sows	0.2	1.5–1.8	55,769–70,000
Mature pigs	0.2	2.9–3.6	111,923–140,000
Poultry			
Laying hens	0.2	1.9	72,308
Nonlaying hens	0.2	3.3	128,077
Turkey	0.2	1.8–2.9	70,000–111,923

Assuming 1 cell contains 0.2 pg toxins (Falconer et al. 1994).
Percent of the TDI allocated to drinking water is 100% (P=1).

ings showed that all cyanobacterial toxins have an impact on animal (both wild and domestic) and aquatic organisms, especially nodularin, which is as potent as microcystin-LR. Attention should also be paid to the occurrence of neurotoxins to which have been attributed many animal poisonings.

The guideline values presented in this study are calculated on the basis of the assumption that the NOAEL is the level at which no histological changes in test animals are observed. In practice, it is not easy to observe this type of effect in animals. In some cases, the effects of the toxins are only recognized when mortality occurs. Therefore, NOAEL for animals can be considered to be the level at which no mortality is observed. Based on this standpoint, TDI is estimated as $0.25\ LD_{50}$. Applying this assumption, the guideline values of microcystin-LR for domestic animal groups can be estimated (Table 15).

Summary

The occurrence of cyanobacterial toxins affects aquatic organisms, terrestrial animals (both wild and domestic), and humans. Detrimental effects have been documented in the scientific literature during the past 50 years. Possible guideline values of some cyanobacterial toxins (microcystins, cylindrospermopsin, and anatoxin-a) are estimated, and they show that children and infants are more susceptible to cyanobacterial toxins than adults. Therefore, particular attention should be paid when cyanobacterial blooms occur, even at relatively low cell counts, to protect children and infants from possible risks. Based on these guideline values and the occurrence of the toxins, it can be concluded that chronic and

72 T. N. Duy et al.

Table 15. Tolerable daily intake (TDI) and guideline value (GV) of microcystin-LR in drinking water for some domestic animal groups.

Animals	TDI	GV (μg/L)	GV (cells/mL)
Cattle			
Dairy cows in milk	1.25	9	44.60×10^6
Dairy cow dry	1.25	14	69.45×10^6
Beef cattle	1.25	13	62.50×10^6
Calves	1.25	3	17.05×10^6
Horses			
Working horse	1.25	14	68.20×10^6
Grazing horse	1.25	21	107.05×10^6
Sheep			
Nursing ewes on dry feed	1.25	7	34.75×10^6
Mature sheep on dry pasture	1.25	9	44.65×10^6
Mature sheep on irrigated pasture	1.25	18	89.30×10^6
Fattening lambs on dry pasture	1.25	28	142.05×10^6
Fattening lambs on irrigated pasture	1.25	57	284.05×10^6
Pigs			
Brood sows	1.25	9–11	$45.45–56.80 \times 10^6$
Mature pigs	1.25	18–23	$90.95–113.65 \times 10^6$
Poultry			
Laying hens	1.25	12	58.65×10^6
Nonlaying hens	1.25	20	101.05×10^6
Turkey	1.25	11–19	$56.80–90.95 \times 10^6$

NOAEL is considered as the level that causes no mortality.
Oral LD_{50} of microcystin-LR is 5 mg/kg mice obtained from subchronic study of Fawell et al. (1993); assuming 0.2 pg microcystin-LR/cell (Falconer et al. 1994).

subchronic exposure to cyanobacterial toxins does occur in some populations, particularly in developing countries where high proportions of the population consume untreated surface water directly, such as pond, ditch, river, or reservoir water. Because wildlife and domestic animals consume a large amount of untreated water daily, they are at higher risk than humans from cyanobacterial toxins. Calculated guideline values in Section X show that a relatively high risk posed by the toxins to these animals is likely to occur, even at low cell densities.

Acknowledgments

P.K.S.L. acknowledges the support of a research grant from the Hong Kong Research Grants Council (City U1057/97M) and the Jackson Memorial Fellowship of Griffith University. We thank also Professor Douglas Park of Louisiana State University for a critical review of the manuscript.

References

Østensvik O, Skulberg OM, Soil NE (1981) Toxicity studies with blue-green algae from Norwegian inland waters. In: Carmichael WW (ed) The Water Environment: Algal Toxins and Health. Plenum Press, New York, pp 315–324.

Adamson RH, Chabner B, Fujiki H (1989) Japan seminar on "Marine Natural Products and Cancer" (meeting report). Jpn J Cancer Res (Gann) 80:1141–1144.

Adelman WJ, Fohlmeister JF, Sasner JJ, Ikawa M (1982) Sodium channels blocked by aphantoxin obtained from the blue-green algae *Aphanizomenon flosaquae*. Toxicon 20:513–516.

Anderson RJ, Luu HA, Chen DZX, Holmes CFB, Kent ML, Le Blanc M, Taylor FJR, Williams DE (1993) Chemical and biological evidence links microcystins to salmon 'netpen liver disease'. Toxicon 31(10):1315–1323.

Aronstam RS, Witkop B (1981) Anatoxin-a interactions with cholinergic synaptic molecules. Proc Natl Acad Sci USA, 78(7):4639–4643.

Astrachan NB, Archer BG, Hilberlink DR (1980) Evaluation of the subacute toxicity and teratogenicity of anatoxin-a. Toxicon 18:684–688.

Bagu J, Sönnichsen FD, Williams DE, Anderson RJ, Sykes BD, Holmes CFB (1995) Comparison of the solution structures of microcystin-LR and motuporin. Nat Struct Biol 2:114–116.

Baker PD (1991) Identification of common noxious cyanobacteria. Report 29. Urban Water Research Association of Australia, Melbourne.

Beasley VR, Cook WO, Dahlem AM, Hooser SB, Lovell AL, Valentine WM (1989) Algal intoxication in livestock and waterfowl. Clin Toxicol Vet Clinics North Am Food Anim Pract 5(2): 345–361.

Bell SG, Codd GA (1996) Detection, analysis, and risk assessment of cyanobacterial toxins. In: Harrison RM, Hester RE (eds) Agricultural Chemicals and the Environment. Royal Society of Chemistry, Information Service, London, pp 109–122.

Benson HJ (1969) Microbiological Application: A Laboratory Manual in General Microbiology, 2nd Ed. Brown, Company, Dubuque, IA.

Berg K, Wyman J, Carmichael WW, Dabholkar A (1988) Isolated rat liver perfusion studies with cyclic heptapeptide toxins of *Microcystis* and *Oscillatoria* (freshwater cyanobacteria). Toxicon 26(9):827–837.

Bishop CT, Anet EFLJ, Gorham PR (1959) Isolation and identification of the fast-death factor in *Microcystis aeruginosa* NRC-1. Can J Biochem Physiol 37:453–471.

Botes DP, Kruger H, Viljoen CC (1982) Isolation and characterization of four toxins from the blue-green algae, *Microcystis aeruginosa*. Toxicon 20:945–954.

Botes DP, Tuinman AA, Wessels PL, Viljoen CC, Kruger H, Williams DH, Santikarn S, Smith RJ, Hammond SJ (1984) The structure of cyanoginosin-LR, a cyclic heptapeptide toxin from the cyanobacterium *Microcystis aeruginosa*. J Chem Soc Perkin Trans 1: 2311–2318.

Botes DP, Wessels PL, Kruger H, Runnegar MTC, Santikarn S, Smith RJ, Barna JCJ, Williams DH (1985) Structure studies on cyanoginosins-LR, -YR, -YA, and YM, peptide toxins from *Microcystis aeruginosa*. J Chem Soc Perkin Trans I:2747–2748.

Bourke ATC, Hawes RB, Neilson A, Stallman ND (1983) An outbreak of hepato-enteritis (the Palm Island mysterious disease) possibly caused by algal intoxication. Toxicon Suppl 3:45–48.

Bourne DG, Jones GJ, Blakeley RL, Jones A, Negri AP, Riddles P (1996) Enzymatic pathway for the bacterial degradation of the cyanobacterial cyclic peptide toxin microcystin-LR. Appl Environ Microbiol 62(11):4086–4094.

Byth S (1980) Palm Island mystery disease. Med J Aust 2:40–42.

Carmichael WW, Biggs DF, Peterson MA (1979) Pharmacology of anatoxin-a, produced by the freshwater cyanophyte *Anabaena flos-aquae* NRC-44-1. Toxicon 17:229–236.

Carmichael WW (1981) Freshwater blue-green algae (cyanobacteria) toxins—review. In: Carmichael WW (ed) The Water Environment: Algal Toxins and Health. Plenum Press, New York, pp 1–13.

Carmichael WW, Gorham PR (1981) The mosaic nature of toxic blooms of cyanobacteria. In: Carmichael WW (ed) The Water Environment: Algal Toxins and Health. Plenum Press, New York, pp 161–172.

Carmichael WW, Mahmood NA (1984) Toxins from freshwater cyanobacteria. In: Ragelis EP (ed) Seafood Toxins. American Chemical Society, Washington, DC, pp 377–389.

Carmichael WW, Jones CLA, Mahmood NA, Theiss WC (1985) Algal toxins and water-based diseases. CRC Crit Rev Environ Control 15(3):275–303.

Carmichael WW (1988) Toxins of freshwater algae. In: Tu AT (ed) Handbook of Natural Toxins: Marine Toxins and Venoms, Vol. 3. Dekker, New York, pp 121–147.

Carmichael WW, Mahmood NA, Hyde EG (1990) Natural toxins from cyanobacteria (blue-green) algae. In: Hall S, Strichartz G (eds) Marine Toxins: Origins, Structure and Molecular Pharmacology. American Chemical Society, Washington, DC, pp 87–106.

Carmichael WW (1992a) A Status Report on Planktonic Cyanobacteria (Blue-Green Algae) and Their Toxins. EPA/600/R-921079. Environmental Monitoring Systems Laboratory, Office of Research and Development, U.S. Environmental Protection Agency, Cincinnati, OH.

Carmichael WW (1992b) Cyanobacteria secondary metabolites—the cyanotoxins. J Appl Bacteriol 72:445–459.

Carmichael WW, Falconer IR (1993) Diseases related to freshwater blue-green algal toxins and control measures. In: Falconer IR (ed) Algal Toxins in Seafood and Drinking Water. Academic Press, London, pp 187–209.

Carmichael WW (1994) The toxins of cyanobacteria. Sci Am 270(1):2–9.

Carr NG, Whitton BA (1982) The Biology of Cyanobacteria. Blackwell.

Chen DZX, Boland MP, Smillie MA, Klix H, Ptak C, Andersen RJ, Holmes CFB (1993) Identification of protein phosphatase inhibitors of the microcystin class in the marine environment. Toxicon 31:1407–1414.

Chiswell R, Smith M, Norris R, Eaglesham G, Shaw G, Seawright A, Moore M (1997) The cyanobacterium, *Cylindrospermopsis raciborskii*, and its related toxin, cylindrospermopsin. Australas J Ecotoxicol 3:7–23.

Codd GA, Bell SG (1985) Eutrophication and toxic cyanobacteria in freshwater. Water Pollut Control 84(2):225–232.

Codd GA, Poon GK (1988) Cyanobacterial toxins. In: Rogers LJ, Gallon JR (eds) Biochemistry of the Algae and Cyanobacteria. Clarendon Press, Oxford, pp 283–297.

Codd GA, Brooks WP, Priestley IM, Poon GK, Bell SG (1989) Production, detection, and quantification of cyanobacterial toxins. Toxic Assess 4:499–511.

Codd GA (1994) Biological aspects of cyanobacterial toxins. In: Steffensen DA, Nicholson BC (eds) Toxic Cyanobacteria, Current Status of Research and Management: International Workshop, 22–26 March 1994, Adelaide, SA. Proceedings, Australian Centre for Water Treatment and Water Quality Research, Salisbury SA.

Codd GA, Steffensen DA, Burch MD, Baker PD (1994) Toxic blooms of cyanobacteria in lake Alexandrina, South Australia—Learning from history. In: Steffensen DA, Nicholson BC (eds) Toxic Cyanobacteria, Current Status of Research and Management: International Workshop, 22–26 March 1994, Adelaide, SA. Proceedings Australian Centre for Water Treatment and Water Quality Research, Salisbury, SA.

Codd GA (1995) Cyanobacterial toxins: occurrence, properties and biological significance. Water Sci Technol 32(4):149–156.

Codd GA, Ward CJ, Bell SG (1997) Cyanobacterial toxins: occurrence, models of action, health effects and exposure routes. Arch Toxicol Suppl 19:399–410.

Collins MD, Gowans CS, Garro F, Estervig D, Swanson T (1981) Temporal association between and algal bloom and mutagenicity in a water reservoir. In: Carmichael WW (ed) The Water Environment: Algal Toxins and Health. Plenum Press, New York pp 271–283.

Cook WO, Beasley VR, Dahlem AM, Delliger JA, Harlin KS, Carmichael WW (1988) Comparison of effects of anatoxin-a(s) and paraoxon, physostigmine and pyridostigmine on mouse brain cholinesterase activity. Toxicon 26(8):750–753.

Craig M, Luu HA, McCready T, Williams DE, Anderson RJ, Holmes CFB (1996) Molecular mechanisms underlying the interaction of motuporin and microcystins with type 1 and 2A protein phosphatase. Biochem Cell Biol 74:569–578.

Dabholkar AS, Carmichael WW (1987) Ultra-structure changes in the mouse liver induced by hepatotoxin from the freshwater cyanobacterium *Microcystis aeruginosa* strain 7820. Toxicon 25:285–292.

Dahlem AM, Hassan AL, Swanson SP, Carmichael WW, Beasley VR (1988) A model system for studying the bioavailability of intestinally administered microcystin-LR, a hepatotoxic peptide from the cyanobacterium, *Microcystis aeruginosa* in the rat. Pharmacol Toxicol 63:1–5.

Davidson FF (1959) Poisoning of wild and domestic animals by a toxic waterblooom of *Nostoc rivulare* Kuetz. J Am Water Works Assoc 51:1277–1287.

Delaney JM, Wilkins RM (1995) Toxicity of microcystin LR isolated from *Microcystis aeruginosa*, against various insect species. Toxicon 33:771–778.

Devlin JP, Edwards OE, Gorham PR, Hunter NR, Pike RK, Stavric B (1977) Anatoxin-a, a toxic alkaloid from *Anabaena flos-aqiae* NRC-44h. Can J Chem 55:1367–1371.

Dillenberg HO, Dehnel MK (1960) Toxic waterbloom in Saskatchewan, 1959. Can Med Assoc J 83:1151–1154.

Drikas M (1994) Control and/or removal of algal toxins. In: Steffensen DA, Nicholson BC (eds) Toxic Cyanobacteria, Current Status of Research and Management: International workshop, 22-26 March 1994, Adelaide, SA. Proceedings Australian Centre for Water Treatment and Water Quality Research, Salisbury, SA.

Edwards C, Beattie KA, Scrimgeour CM, Codd GA (1992) Identification of anatoxin-a in benthic cyanobacteria (blue-green algae) and in associated dog poisoning at Loch Insh, Scotland. Toxicon 30:1165–1175.

Eloff JN, Van der Westhuizen AJ (1981) Toxicological studies on *Microcystis*. In: Carmichael WW (ed) The Water Environment: Algal Toxins and Health. Plenum Press, New York, pp 343–363.

El Saadi O, Cameron AS (1993) Illness associated with blue-green algae. Med J Aust 158:792–793.

Eriksson JE, Meriluoto JAO, Lindholm T (1989) Accumulation of a peptide toxin from the cyanobacterium *Oscillatoria agardhii* in the freshwater mussel *Anadonta cygnea*. Hydrobiologia 183:211–216.

Eriksson JE, Gronberg L, Nygard S, Slotte JP, Meriluoto JAO (1990a) Hepatocellular uptake of ₃H-dihydromicrocystin-LR, a cyclic peptide toxin. Biochem Biophys Acta 1025:60–66.

Eriksson JE, Toivola D, Meriluoto JAO, Karaki H, Han YG, Hartshorne D (1990b) Hepatocyte deformation induced by cyanobacterial toxins reflect inhibition of protein phosphatases. Biochem Biophys Res Commun 173(3):1347–1353.

Falconer IR, Jackson ARB, Langley J, Runnegar MT (1981) Liver pathology in mice in poisoning by the blue-green alga *Microcystis aeruginosa*. Aust J Biol Sci 34:179–187.

Falconer IR, Runnegar MTC, Jackson ARB, McInnes A (1983) The occurrence and consequences of blooms of the toxic blue-green alga *Microcystis aeruginosa* in Eastern Australia. Toxicon Suppl 3:119–121.

Falconer IR, Buckley T, Runnegar MTC (1986) Biological half-life, organ distribution and excretion of ^{125}I-labelled toxic peptide from blue-green alga *Microcystis aeruginosa*. Aust J Biol Sci 39:17–21.

Falconer IR, Smith JV, Jackson ARB, Jones A, Runnergar MTC (1988) Oral toxicity of a bloom of the cyanobacterium *Microcystis aeruginosa* administered to mice over periods up to 1 year. J Toxicol Environ Health 24:291–305.

Falconer IR, Buckley TH (1989) Tumour promotion by *Microcystis* sp., a blue-green alga occurring in water supplies. Med J Aust 150:351.

Falconer IR, Runnegar M, Buckley T, Huyn V, Bradshaw P (1989) Using activated carbon to remove toxicity from drinking water containing cyanobacterial blooms. J Am Water Works Assoc 81(2):102–105.

Falconer IR (1991) Tumour promotion and liver injury caused by oral consumption of cyanobacteria. Environ Toxicol Water Qual 6:177–184.

Falconer IR, Yeung DSK (1992) Cytoskeletal changes in hepatocytes induced by *Microcystis* toxins and their relation to hyperphosphorylation of cell proteins. Chem Biol Interact 81:181–196.

Falconer IR, Dornbusch M, Moran G, Yeung SK (1992) Effect of the cyanobacterial (blue-green algal) toxins from *Microcystis aeruginosa* on isolated enterocystes from the chicken small intestine. Toxicon 30(7):790–793.

Falconer IR (1993a) Measurement of toxins from blue-green algal in water and foodstuffs. In: Falconer IA (ed) Algal Toxins in Seafood and Drinking Water. Academic Press, London, pp 165–175.

Falconer IR (1993b) Mechanism of toxicity of cyclic peptide toxins from blue-green algae. In: Falconer IA (ed) Algal Toxins in Seafood and Drinking Water Academic Press, London, pp 177–186.

Falconer IR (1994a) Health implications of cyanobacterial (blue-green algal) toxins. In: Steffensen DA, Nicholson BC (eds) Toxic Cyanobacteria, Current Status of Research and Management: International Workshop, 22–26 March 1994, Adelaide, SA. Proceedings, Australian Centre for Water Treatment and Water Quality Research, Salisbury, SA, pp 61–65.

Falconer IR (1994b) Health problems from exposure to cyanobacteria and proposed safety guidelines for drinking and recreational water. In: Codd GA, Jefferies TM, Keevil CW, Potter E (eds) Detection Methods for Cyanobacterial Toxins. Royal Society of Chemistry, London, pp 3–10.

Falconer IR, Burch MD, Steffensen AD, Choice M, Coverdale BR (1994) Toxicity of the blue-green alga (cyanobacterium) *Microcystis aeruginosa* in drinking water to growing pigs, as an animal model for human injury and risk assessment. Environ Toxicol Water Qual 9:131–139.

Fawell JK, James CP, James HA (1993) Toxins from blue-green algae: toxicological assessment of microcystin-LR and a method for its determination in water. Foundation for Water Research, Marlow, Bucks.

Fitzgeorge RB, Clark SA, Keevil CW (1994) Routes of intoxication. In: Codd GA, Jefferies TM, Keevil CW, Potter C (eds) Detection Methods for Cyanobacterial Toxins. Royal Society of Chemistry, London, pp 69–74.

Gaete V, Canelo E, Lagos N, Zambrano F (1994) Inhibitory effects of *Microcystis aeruginosa* toxin on ion pumps of the gill of freshwater fish. Toxicon 32(1):121–127.

Galey FD, Beasley VR, Carmichael WW, Hooser SB, Haschek-Hock WM, Lovell RA, Poppenga RH, Knight MW (1986) Blue-green algae hepatotoxicosis in cattle. Proceedings of the American Chemical Society 8th Rocky Mountain Regional Meeting, Denver, CO, June 8–12 (abstract).

Galey FD, Beasley VR, Carmicjael WW, Kleppe G, Hooser SB, Haschek WM (1987) Blue-green algae (*Microcystis aeruginosa*) hepatotoxicosis in dairy cows. Am J Vet Res 48:1415–1420.

Gallon JP, Chit KN, Brown EG (1990) Biosynthesis of the tropane-related cyanobacterial toxin anatoxin-a: role of ornithine decarboxylase. Phytochemistry (Oxf) 29(4):1107–1111.

Gathercole PS, Thiel PG (1987) Liquid chromatographic determination of the cyanoginosins, toxins produced by the cyanobacterium *Microcystin aeruginosa*. J Chromatogr 408:435–440.

Gorham PR (1964) Toxin algae. In: Jackson DF (ed) Algae and Man. Plenum Press, New York, pp 307–336.

Gorham PR, Carmichael WW (1988) Hazards of freshwater blue-green algae (cyanobacteria). In: Lembi CA, Waaland JR (eds) Algae and Human Affairs. Cambridge University Press, Cambridge, pp 403–431.

Harada KI, Ogawa K, Matsuura K, Nagai H, Murata H, Suzuki M, Itezono Y, Nakayama N, Shirai M, Nakano M (1991) Isolation of two toxic heptapeptide microcystins from an axenic strain of *Microcystis aeruginosa*, K-139. Toxicon 29:479–489.

Harada KI, Ohtani I, Iwamoto K, Suzuki M, Watanabe MF, Watanabe M, Terao K (1994a) Isolation of cylindrospermopsin from a cyanobacterium *Umezakia natans* and its screening method. Toxicon 32(1):73–84.

Harada KI, Suzuki M, Watanabe MF (1994b) Structure analysis of cyanobacterial toxins. In: Codd GA, Jefferies TM, Keevil, CW, Potter E (eds) Detection Methods for Cyanobacterial Toxins. The Royal Society of Chemistry, London, pp 103–148.

Hawkins PR, Runnergar MTC, Jackson ARB, Falconer IR (1985) Severe hepatotoxicity caused by the tropical cyanobacterium (blue-green alga) *Cylindromopsis raciborskii* (Woloszynska) Seenaya and Subba Raju isolated from a domestic water supply reservoir. Appl Environ Microbiol 50(50):1292–1295.

Hawkins PR, Chandrasena NR, Jones GJ, Humpage AR, Falconer IR (1997) Isolation and toxicity of *Cylindrospermopsis raciborskii* from an ornamental lake. Toxicon 35:341–346.

Hayman J (1992) Beyond the Barcoo—probable human tropical cyanobacterial poisoning in outback Australia. Med J Aust 157:794–796.

Henriksen P, Carmichael WW, An J, Moestrup Ø (1997) Detection of an anatoxin-a(s) like anticholinesterase in natural blooms and cultures of cyanobacteria/blue-green algae from Danish lakes and in the stomach contents of poisoned birds. Toxicon 35(6):901–913.

Himberg K, Keijola AM, Hiisvirta L, Pyysalo H, Sivonen K (1989) The effect of water treatment processes on the removal of hepatotoxins from *Microcystis* and *Oscillatoria* cyanobacteria: a laboratory study. Water Res 23(8):979–984.

Hoffman JRH (1976) Removal of *Microcystis* toxins in water purification process. Water SA (Pretoria) 2:58–60.

Holmes CFB, McGready TL, Craig M, Lambert TW, Hrudney SE (1994) A sensitive bioscreen for detection of cyclic peptide toxins of the microcystin class. In: Codd GA, Jefferies TM, Keevil CW, Potter E (eds) Detection Methods for Cyanobacterial Toxins. Royal Society of Chemistry, London, pp 85–89.

Honkanen RE, Zwiller J, Moore RE, Daily SL, Khatra BS, Dukelow M, Boynton AL (1990) Characterization of microcystin-LR, a potent inhibitor of type 1 and type 2A protein phosphatases. J Biol Chem 256:19401–19404.

Hooser SB, Kuhlenschmidt MS, Dahlem AM, Beasley VR, Carmichael WW, Haschek WM (1991) Uptake and subcellular localization of tritiated dihydro-microcystin-LR in rat liver. Toxicon 29(6):589–601.

Hrudney SE, Kenefick SL, Lambert TW, Kotak BG, Prepas EE, Holmes CFB (1994a) Sources of uncertainty in assessing the health risk of cyanobacterial blooms in drinking water supplies. In: Codd GA, Jefferies TM, Keevil CW, Potter E (eds) Detection Methods for Cyanobacterial Toxins. Royal Society of Chemistry, London, pp 122–130.

Hrudney SE, Lambert TW, Kenefick SL (1994b) Health risk assessment of microcystins in drinking water supplies. In: Toxic Cyanobacteria: a Global Perspective Symposium, 28 March, 1994. Australian Centre for Water Quality Research, Adelaide, SA, pp 7–16.

Huber CS (1972) The crystal structure and absolute configuration of 2,9-diacetyl-9-azabicyclo[4,2,1] non-2,3-ene. Acta Crystallog 238:2577–2582.

Humpage AR, Rositano J, Bretag AH, Brown R, Baker PD, Nicholson BC, Steffensen DA (1994) Paralytic shellfish poisons from Australian cyanobacteria blooms. Aust J Mar Fresh Water Res 45:761–771.

IARC (International Agency for Research on Cancer) (1987) IARC monographs on the evaluation of carcinogenic risks to humans. Suppl. 7. Overall evaluations of carcinogenicity: an updating of IARC monographs, Vols. 1–42. IARC, Lyon.

Ilyaletdinova SG, Dubitskiy AM (1972) A method for determining the toxicity of blue-green algae with respect to mosquito larvae. Izv AN Kaz SSR Ser Biol 2:39–42.

International Code of Botanical Nomenclature (1972) International Association for Plant Taxonomy, Utrecht.

Jackim E, Gentile J (1968) Toxins of a blue-greeen alga: similarity to saxitoxin. Science 162:915–916.

Jackson ARB, McInnes A, Falconer IR, Runnegar MTC (1984) Clinical and pathological changes in sheep experimentally poisoned by blue-green algae *Microcystis aeruginosa*. Vet Pathol 21:102–113.

Jann K, Jann B (1984) Structure and biosynthesis of O-antigens. In: Rietschel ET (ed) Handbook of Endotoxin, Vol. 1. Chemistry of Endotoxin. Elsevier, Amsterdam, pp 138–186.

Jochimsen EM, Carmichael WW, An J, Cardo DM, Cookson ST, Holmes CEM, Antunes MB de C, Filho de M DA, Lyra TM, Barreto VST, Azevedo SMFO, Jarvis WR (1998)

Liver failure and death after exposure to microcystins at a hemodialysis centre in Brazil. N Eng J Med 338(13):873–878.

Johnstone P (1994) Cyanobacteria in Australia—an overview. In: Steffensen DA, Nicholson BC (eds) Toxic Cyanobacteria, Current Status of Research and Management: International Workshop, 22–26 March 1994, Adelaide, SA. Proceedings, Australian Centre for Water Treatment and Water Quality Research, Salisbury, SA.

Jones GJ (1990) Biodegradation and removal of cyanobacterial toxins in natural waters. In: Proceedings of Sydney Water Board Blue-green Algae Seminar, Sydney, pp 33–36.

Jones GJ (1994) Cyanobacterial bloom: a 'natural' feature of impounded Australian water? In: Steffensen DA, Nicholson BC (eds) Toxic Cyanobacteria, Current Status of Research and Management: International Workshop, 22–26 March 1994, Adelaide, SA. Proceedings, Australian Centre for Water Treatment and Water Quality Research, Salisbury SA.

Jones GJ, Orr PT (1994) Release and degradation of microcystin following algicide treatment of a *Microcystis aeruginosa* bloom in a recreational lake, as determined by HPLC and protein phosphatase inhibition assay. Water Res 28(4):871–876.

Jones GJ, Falconer IR, Wilkins RM (1995) Persistence of cyclic peptide toxins in dried *Microcystis aeruginosa* crusts from Lake Mokoan, Australia. Environ Toxicol Water Qual 10:19–24.

Kaya K (1996) Toxicology of microcystins. In: Watanabe M, Harada KI, Carmichael WW, Fujiki H (eds) Toxic Microcystis. CRC Press, Boca Raton, UK, pp 175–202.

Keijola AM, Himberg K, Esala AL, Sivonen K, Hiisvirta L (1988) Removal of cyanobacterial toxins in water treatment processes: laboratory and pilot scale experiments. Toxic Assess 3:643–656.

Kenefick SL, Hrudey SE, Peterson HG, Prepas EE (1993) Toxin release from *Microcystis aeruginosa* after chemical treatment. Water Sci Technol 27(3-4):433–440.

Kent ML (1990) Netpen liver disease (NLD) of salmonid fishes reared in sea water: species susceptibility, recovery, and probable cause. Dis Aquat Org 8:21–28.

Kent ML, Meyers MS, Hinton DE, Eaton WD, Elston RA (1988) Suspected toxicopathic hepatic necrosis and megalocytosis in pen-reared Atlantic salmon *Salmo salar* in Puget Sound, Washington, USA. Dis Aquat Org 4:91–100.

Kirpenko YA, Sirenko LA, Kirpenko NI (1981) Some aspects concerning remote after-effects of blue-green algae toxins impact on warm-blooded animal. In: Carmichael WW (ed) The Water Environment: Algal Toxins and Health. Plenum Press, New York, pp 257–259.

Kiviranta J, Sivonen K, Lahti K, Luukkainen R, Niemel™ SI (1991) Production and biodegradation of cyanobacterial toxins—a laboratory study. Arch Hydrobiol 121(3):281–294.

Kofuji P, Aracava Y, Swanson KL, Aronstam RS, Rapoport H, Albuquerque EX (1990) Activation and blockade of the acetylcholine receptor-ion channel by the agonists (+)-anatoxin-a, the *N*-methyl derivative and the errant enantiomer. J Pharmacol Exp Ther 252:517–525.

Koskinen AMP, Rapoport H (1985) Synthetic and conformational study on anatoxin-a: a potent acetylcholine agonist. J Med Chem 28:1301.

Kotak BG, Semalulu S, Fritz DL, Prepas EE, Hrudney SE, Coppock RW (1996a) Hepatic and renal pathology of intraperitoneally administered microcystin-LR in rainbow trout (*Oncorhynchus mykiss*). Toxicon 34(5):517–525.

Kotak BG, Zurawell RW, Prepas EE, Holmes CFB (1996b) Microcystin-LR concentration in aquatic food web compartments from lakes of varying trophic status. Can J Fish Aquat Sci 53:1974–1985.

Krishnamurthy T, Szafraniec L, Hunt DF, Shabanowitz J, Yates JR, Hauer CR, Carmichael WW, Skulberg OM, Codd GA, Missler S (1989) Structural characterization of toxic cyclic peptides from blue-green algae by tandem mass spectrometry. Proc Natl Acad Sci USA 86:770–774.

Kuiper-Goodman T, Gupta S, Combley H, Thomas BH (1994) Microcystins in drinking water: risk assessment and derivation of a possible guidance value for drinking water. In: Steffensen DA, Nicholson BC (eds) Toxic Cyanobacteria, Current Status of Research and Management: International Workshop, 22–26 March 1994, Adelaide, SA. Proceedings, Australian Centre for Water Treatment and Water Quality Research, Salisbury, SA, pp 67–73.

Lahti K (1997) Cyanobacterial hepatotoxins and drinking water supplies—aspects of monitoring and potential health risks. Monographs of Boreal Environment Research No. 4. Finnish Environment Institute, Finland.

Lahti K, Rapala J, Fardig M, Niemela M, Sivonen K (1997) Persistence of cyanobacterial hepatotoxin, microcystin-LR in particulate material and dissolved in lake water. Water Res 31(5):1005–1012.

Lam AKY, Fedorak PM, Prepas EE (1995) Biotransformation of cyanobacterial hepato-toxin microcystin-LR, as determined by HPLC and protein phosphatase bioassay. Environ Sci Technol 29:242–246.

Lambert TW (1993) Removal of microcystin-LR toxin from drinking water. Master's thesis. University of Alberta, Edmonton, Alberta, Canada.

Lambert TW, Boland M, Holmes CFB, Hrudney SE (1994) Quantitation of the microcystin hepatotoxins in water at environmentally relevant concentrations with the protein phosphatase bioassay. Environ Sci Technol 28:753–755.

Lawton LA, Codd GA (1991) Cyanobacterial (blue-green algal) toxins and their significance in UK and European Waters. J Inst Water Environ Manage 5:460–465.

Lembi CA, Waaland JR (1988) Algae and Human Affairs. Cambridge University Press, Cambridge.

Lippy EC, Erb J (1976) Gastrointestinal illness at Sewickley, Pa. J Am Water Works Assoc 68:606–610.

Lukac M, Aegerter R (1993) Influence of trace metals on growth and toxin production of *Microcystis aeruginosa*. Toxicon 31(3):293–305.

Luukkainen R, Sivonen D, Namikoshi M, F™rdig M, Rinehart KL, Niemel™ SI (1993) Isolation and identification of eight microcystins from thirteen *Oscillatoria agardhii* strains and structure of a new microcystin. Appl Environ Microbiol 59(7):2204–2209.

MacKintosh C, Beattie KA, Klumpp S, Cohen P, Codd GA (1990) Cyanobacterial microcystin-LR is a potent and specific inhibitor of protein phosphatases 1 and 2A from both mammals and higher plants. FEBS Lett 264:187–192.

Mahmood WA, Carmichael WW (1986) Paralytic shellfish poisons produced by the freshwater cyanobacterium *Aphanizomenon flos-aquae* NH-5. Toxicon 24(2):175–186.

Mahmood WA, Carmichael WW (1987) Anatoxin-a(s), an anticholinesterase from the cyanobacterium *Anabaena flos-aquae* NRC-525-17. Toxicon 25(11):1211–1227.

Martin C, Codd GA, Sigelman HW, Weckesser L (1989) Lipopolysaccharides and polysaccharides of the cell envelope of toxic *Microcystis aeruginosa* strain. Arch Microbiol 152:90–94.

Matsunaga S, Moore RE, Niemczura WP, Carmichael WW (1989) Anatoxin-a(s), a potent anticholinesterase from *Anabaena flos-aquae*. J Am Chem Soc 111:8021–8023.

Matsushima R, Yoshizawa S, Watanabe MF, Harada KI, Furusawa M, Carmichael WW, Fujiki H (1990) In vitro and in vivo effects of protein phosphatase inhibitors, microcystins and nodularin, on mouse skin and fibroblasts. Biochem Biophys Res Commun 172:867–874.

Mayer H, Weckesser J (1984) 'Unusual' lipid A's: structures, taxonomical relevance and potential value for endotoxin research. In: Rietschel ET (ed) Handbook of Endotoxin, Vol. 1. Chemistry of Endotoxin. Elsevier, Amsterdam, pp 221–247.

Meriluoto JAO, Eriksson JE (1988) Rapid analysis of peptide toxin in cyanobacteria. J Chromatogr 438:93–99.

Meriluoto JAO, Sandstorm A, Eriksson JE, Remaud G, Craig AG, Chattopadhyaya J (1989) Structure and toxicity of a peptide hepatotoxin from the cyanobacterium *Oscillatoria agardhii*. Toxicon 27:1021–1034.

Meriluoto JAO, Nygård SE, Dahlem AM, Eriksson JE (1990) Synthesis, organotropism and hepatocellular uptake of two tritium-labeled epimers of dihydromicrocystin-LR, a cyanobacterial peptide toxin analog. Toxicon 28(12):1439–1446.

Moore RE, Ohtani L, Moore BS, DeKoning CB, Yoshida WY, Runnegar MTC, Carmichael WW (1993) Cyanobacterial toxins. Gazz Chim Ital 123:329–336.

Namikoshi M, Rinehart KL, Sakai R, Sivonen K, Carmichael WW (1990) Structures of three new cyclic heptapeptide heptatotoxins produced by the cyanobacterium (blue-green alga) *Nostroc* sp. strain 152. J Org Chem 55:6135–6139.

Namikoshi M, Sivonen K, Evans WR, Carmichael WW, Rouhiainen L, Luukkainen, Rinehart KL (1992) Structures of three new homotyrosine-containing microcystins and a new homophenylaine variant from *Anabaena* sp. strain 66. Chem Res Toxicol 5:661–666.

Negri AP, Jones GJ, Blackburn SI, Oshima Y, Onodera H (1994) Effect of culture and bloom development and of sample storage on paralytic shellfish poisons in the cyanobacterium *Anabaena circinalis*. J Phycol 33:26–35.

Nicholson BC, Rositano J, Burch MD (1994) Destruction of cyanobacterial peptide hepatotoxins by chlorine and chloramine. Water Res 28(6):1297–1303.

Nishiwaki-Matsushima R, Nishiwaki S, Ohta T, Yoshizawa S, Suganuma M, Harada K, Watanabe MF, Fujiki H (1991) Structure-function relationships of microcystins, liver tumor promoters, in interaction with protein phosphatase. Jpn J Cancer Res 82:993–996.

Nishiwaki-Matsushima R, Ohta T, Nishiwaki S, Suganuma M, Kohyama K, Ishikawa T, Carmichael WW, Fujiki H (1992) Liver tumour promotion by the cyanobacterial cyclic peptide toxin microcystin-LR. J Cancer Res Clin Oncol 118:420–424.

Ohtani I, Moore RE (1992) Cylindrospermopsin: a potent hepatotoxin from the blue-green alga *Cylindrospermopsis raciborskii*. J Am Chem Soc 114:7941–7942.

Onodera H, Oshima Y, Henriksen P, Yasumoto T (1997) Confirmation of anatoxin-a(s) in the cyanobacterium *Anabaena lemmermannii*, as the cause of bird kills in Danish lakes. Toxicon 35(11):1645–1648.

Penaloza R, Rojas M, Vila L, Zambrano F (1990) Toxicity of a soluble peptide from *Microcystis* sp. to zooplankton and fish. Freshwater Biol 24:233–240.

Prepas EE, Kotak BG, Campbell LM, Evans JC, Hrudney SE, Holmes CFB (1997) Accumulation and elimination of cyanobacterial hepatotoxins by the freshwater clam *Anodonta grandis simpsoniana*. Can J Fish Aquat Sci 54:41–46.

Queensland Water Quality Task Force (QWQTF) (1992) Freshwater algal blooms in Queensland QWQTF, Queensland, Australia.

Rabergh CML, Bylund G, Eriksson JE (1991) Histopathological effects of microcystin-LR, a cyclic peptide toxin from the cyanobacterium (blue-green alga) *Microcystis aeruginosa*, on common carp (*Cyprinus carpio* L.). Aquat Toxicol 20:131–146.

Rai AN (1990) Handbook of Symbiotic Cyanobacteria. CRC Press, Boca Raton.

Rapala J, Lahti K, Sivonen K, Niemelä SI (1994) Biodegradability and absorption on lake sediment of cyanobacterial hepatotoxins and anatoxin-a. Lett Appl Microbiol 19:423–428.

Repavich WM, Sonzogni WC, Standridge JH, Wedepohl RE, Meisner L (1990) Cyanobacteria (blue-green algae) in Wisconsin waters: acute and chronic toxicity. Water Res 24(2):225–231.

Ressom R, Soong FS, Filzgerald J, Turczynowicz L, El Saadi O, Roder D, Maynard T, Falconer I (1994) Health Effects of Toxic Cyanobacterial (Blue-Green Algae). National Health and Medical Research Council (NHMRC), Australia, Canberra.

Rinehart KL, Harada KI, Namikoshi M, Chen C, Harvis CA, Munro MHG, Blunt JW, Mulligan PE, Beasley VR, Dahlem AM, Carmichael WW (1988) Nodularin, microcystin, and the configuration of Adda. J Am Chem Soc 110: 8557–8558.

Robinson NA, Miura GA, Matson CF, Dinterman RE, Pace JG (1989) Characterisation of chemically tritiated microcystin-LR and its distribution in mice. Toxicon 27(9):1035–1042.

Runnegar MT, Falconer IR (1981) Isolation, characterisation, and pathology of the toxin from the blue-green alga *Microcystis aeruginosa*. In: Carmichael WW (ed) The Water Environment: Algal Toxins and Health. Plenum Press, New York, pp 325–342.

Runnegar MTC, Falconer IR, Silver J (1981) Deformation of isolated rat hepatocytes by a peptide hepatoxin from the blue-green alga *Microcystis aeruginosa*. Arch Pharmacol 317:268–272.

Runnegar MTC, Falconer IR (1982) The *in vivo* and *in vitro* biological effects of the peptide hepatotoxin from the blue-green alga *Microcystis aeruginosa*. S Afr J Sci 78:363–366.

Runnegar MTC, Falconer IR, Jackson ARB, McInnes A (1983) Toxin production by *Microcystis aeruginosa* cultures. Toxicon Suppl 3:377–380.

Runnegar MTC, Falconer IR (1986) Effect of toxin from the cyanobacterium *Microcystis aeruginosa* on ultra-structural morphology and actin polymerization in isolated hepatocytes. Toxicon 24:109–115.

Runnegar MTC, Jackson ARB, Falconer IR (1988) Toxicity of the cyanobacterium *Nodularia spumigena* Mertens. Toxicon 26(2):143–151.

Sadler R (1994) Report on Cyanobacteria in Water. Government Chemical Laboratory, Coopers Plains, Queensland, Australia.

Sasner JJ, Ikawa M, Foxall TL (1984) Studies on *Aphanizomenon and Microcystis* toxins. In: Ragelis EP (ed) Seafood Toxins. American Chemical Society, Washington, DC, pp 391–406.

Sawyer PJ, Gentile JH, Sasner JJ Jr (1968) Demonstration of a toxin from *Aphanizomenon flos-aquae* (L.) Ralfs. Can J Microbiol 14:1199–1204.

Shaw GR, Sukenik A, Livne A, Chiswell RK, Smith MJ, Seawright AA, Norris RL, Eaglesham GK, Moore MR (1999) Blooms of the cylindrospermopsin containing cyanobacterium, *Aphanizomenon ovalisporum* (Forti) in newly constructed lakes, Queensland, Australia. Environ Toxicol, 14:167–177.

Siegelman HW, Adams WH, Stoner RD, Slatkin DN (1984) Toxins of *Microcystis aeruginosa* and their hematological and histopathological effects. In: Ragelis EP (ed) Seafood Toxins. American Chemical Society, Washington, DC, pp 407–413.

Sivonen K, Himberg K, Luukkainen R, Niemelä SI, Poon GK, Codd GA (1989) Preliminary characterization of neurotoxic cyanobacteria blooms and strains from Finland. Toxic Assess 4:339–352.

Sivonen K (1990) Effects of light, temperature, nitrate, orthophosphate and bacteria on growth of and hepatotoxin production by *Oscillatoria agardhii* strains. Appl Environ Microbiol 56(9):2658–2666.

Sivonen K, Carmichael WW, Namikoshi M, Rinehart KL, Dahlem AM, Niemelä SI (1991) Isolation and characterization of heptatotoxic microcystin homologs from the filamentous freshwater cyanobacterium *Nostoc* sp. strain 152. Appl Environ Microbiol 56:2650–2657.

Skulberg OM, Carmichael WW, Anderson RA, Matsunaga S, Moore RE, Skulberg R (1992) Investigations of a neurotoxic oscillatorialean stain (Cyanophyceae) and its toxin. Isolation and characterization of homoanatoxin-a. Environ Toxicol Chem 11:321–329.

Skulberg OM, Carmichael WW, Codd GA, Skulberg R (1993) Taxonomy of toxic Cyanophyceae (Cyanobacteria). In: Falconer IA (ed) Algal Toxins in Seafood and Drinking Water. Academic Press, London, pp 145–163.

Smith RA, Lewis D (1987) A rapid analysis of water for anatoxin-a, the unstable toxic alkaloid from *Anabaena flos-aquae*, the stable non-toxic alkaloid left after bioreduction and a related amine which may be nature's precursor to anatoxin-a. Vet HumToxicol 29(2):153–154.

Soong FS, Maynard E, Kirke K, Luke C (1992) Illness associated with blue-green algae. Med J Aust 156(1):67.

Spivak CE, Witkop B, Albuquerque EX (1980) Anatoxin-a: a novel, potent agonist at the nicotinic receptor. Mol Pharmacol 18:384–394.

Steffensen DA, Nicholson BC (eds) (1994) Toxic Cyanobacteria, Current Status of Research and Management: International Workshop, 22–26 March 1994, Adelaide, SA. Proceedings, Australian Centre for Water Treatment and Water Quality Research, Salisbury, SA.

Stevens DK, Krieger RI (1991) Stability studies on the cyanobacterial nicotinic alkaloid anatoxin-a. Toxicon 29(2):167–179.

Stotts RR, Namikoshi M, Haschek WM, Rinehart KL, Carmichael WW, Dahlem AM, Beasley V (1993) Structural modifications imparting reduced toxicity in microcystins from *Microcystis* spp. Toxicon 31(6):783–789.

Sykora JL, Keleti G (1981) Cyanobacteria and endotoxins in drinking water supplies. In: The Water Environment: Algal Toxins and Health. Plenum Press, New York, pp 285–302.

Terao K, Ohmori S, Igarashi K, Ohtani I, Watanabe MF, Harada KI, Ito E, Watanabe M (1994) Electron microscopic studies on experimental poisoning in mice induced by cylindrospermopsin isolated from blue-green alga *Umezakia natans*. Toxicon 32(7):833–843.

Theiss WC, Carmichael WW, Wyman J, Bruner R (1988) Blood pressure and hepatocellular effects of the cyclic heptapeptide toxin produced by *Microcystis aeruginosa* strain PCC-7820. Toxicon 26:603–613.

Tisdale ES (1931) The 1930-1931 drought and its effect upon public water supply. Am J Public Health 21:1203–1218.

Tsuji K, Watanuki T, Kondo F, Watanabe MF, Suzuki S, Nakaza H, Suzuki M, Uchida H, Harad KI (1995) Stability of microcystins from Cyanobacteria. II. Effect of UV light on decomposition and isomerization. Toxicon 33:1619–1631.

Turner PC, Gammie AJ, Hollinrake K, Codd GA (1990) Pneumonia associated with Cyanobacteria. Br Med J 300:1440–1441.

Ueno Y, Nagata S, Tsutsumi T, Hasegawa A, Watanabe MF, Park HD, Chen GC, Yu SH (1996) Detection of microcystins, a blue-green algal hepatotoxin in drinking water sampled in Haimen and Fusui, endemic areas of primary liver cancer in China, by highly sensitive immunoassay. Carcinogenesis (Oxf) 17(6):1317–1321.

Vackeria D, Codd GA, Bell SG, Beattie KA, Priestley IM (1985) Toxicity and extrachromosomal DNA in strains of the cyanobacterium *Microcystis aeruginosa*. FEMS Microbiol Lett 29:69–72.

Van den Hoek C, Mann DG, Jahns HM (1995) Algae: An Introduction to Phycology. Cambridge University Press, Cambridge.

Van der Westhuizen AJ, Eloff JN (1983) Effect of culture age and pH of culture medium on the growth and toxicity of the blue-green alga *Microcystis aeruginosa*. Z Pflanzenphysiol 110:157–163.

Van der Westhuizen AJ, Eloff JN (1985) Effect of temperature and light on the toxicity and growth of the blue-green alga *Microcystis aeruginosa* (UV 600). Planta (Heidello) 163:55–59

Vasconcelos VM (1995) Uptake and depuration of the heptapeptide toxin microcystin-LR in *Mytilus galloprovincialis*. Aquat Toxicol 32:227–237.

Veldee MV (1931) An epidemiological study of suspected water-borne gastroenteritis. Am J Public Health 21(9):1227–1235.

Wannemacher RW, Bunner DL, Dinterman RE (1987) Comparison of toxicity and absorption of algal toxins and microcystins after dermal exposure in guinea pigs. In: Gopalakvishakone P, Tan CK (eds) Progress in Venom and Toxin Research. Faculty of Medicine, National University of Singapore, pp 718–723.

Wardlaw AC (1985) Practical Statistics for Experimental Biologists. Wiley, New York.

Watanabe MF, Oishi S (1985) Effects of environmental factors on toxicology of a cyanobacterium *(Microcystis aeruginosa)* under culture conditions. Appl Environ Microbiol 49(5):1342–1344.

Watanabe MF, Kaya K, Takamura N (1992) Fate of the toxic cyclic heptapeptides, the microcystins, from blooms of *Microcystis* (cyanobacteria) in a hypertrophic lake. J Phycol 28:761–767.

William DE, Kent ML, Anderson RJ, Klix H, Holmes CFB (1995) Tissue distribution and clearance of tritium-labeled dihydromicrocystin-LR epimers administered to Atlantic salmon via intraperitoneal injection. Toxicon 33(2):125–131.

Yoshida T, Makita Y, Nagata S, Tsutsumi T, Yoshida F, Sekijima M, Tamura SI, Ueno Y (1997) Acute oral toxicity of microcystin-LR, a cyanobacterial hepatotoxin, in mice. Nat Toxins 5:91–95.

Yoshizawa S, Matsushi R, Watanabe MF, Harada KI, Ichihara A, Carmichael WW, Fujiki H (1990) Inhibition of protein phosphatases by microcystin and nodularin associated with hepatotoxicity. J Cancer Res Clin Oncol 116:609–614.

Yu SZ (1989) Drinking water and primary liver cancer. In: Tang ZY, Wu MC, Xia SS (eds) Primary Liver Cancer. Springer, Berlin, pp 30–37.

Yu SH (1994) Blue-green algae and liver cancer, In: Steffensen DA, Nicholson BC (eds) Toxic Cyanobacteria, Current Status of Research and Management: International

Workshop, 22–26 March 1994, Adelaide, SA. Proceedings, Australian Centre for Water Treatment and Water Quality Research, Salisbury, SA, pp 22–26.

Yu SZ (1995) Primary prevention of hepatocelllular carcinoma. J Gastroenterol Hepatol 10:674–682.

Zambrano F, Canelo E (1996) Effects of microcystin-LR on the partical reactions of the Na^+-K^+ pump of the gill of carp (*Cyprinus carpio* linneaus). Toxicon 34(4):451–458.

Zilberg B (1966) Gastroenteritis in Salisbury European children—a five-year-study. Cent Afr J Med 12(9):164–168.

Manuscript received December 1, 1998, accepted January 8, 1999.

Index